Springer Series in Advanced Manufacturing

Series Editor

Professor D.T. Pham
Manufacturing Engineering Centre
Cardiff University
Queen's Building
Newport Road
Cardiff CF24 3AA
UK

Nirav N. Chokshi • Duncan C. McFarlane

A Distributed Coordination Approach to Reconfigurable Process Control

 Springer

Nirav N. Chokshi, PhD
Duncan C. McFarlane, PhD

Institute for Manufacturing (IfM)
Department of Engineering
University of Cambridge
Cambridge CB2 1RX
UK

ISBN 978-1-84996-717-4   e-ISBN 978-1-84800-060-5

DOI 10.1007/978-1-84800-060-5

Springer Series in Advanced Manufacturing ISSN 1860-5168

British Library Cataloguing in Publication Data
A catalogue record for this book is available from the British Library

Preface

The field of Reconfigurable Process Control – the control of process operations requiring a high degree of reconfigurability – is relatively new although its roots go back almost thirty years in the early work of Morari and his co-workers. Until recently, research in this direction has been at a relatively low level due to the industry's preferred focus on mass production and on reducing production costs. But with recent trends in the process industry – increased competition and globalisation that require customisation in products and processes – the notion of reconfigurability is again on the agenda of industrial practitioners, and is thus deserving of research effort.

The problem of reconfigurability in process operations can be examined in a number of different ways. The research presented in this monograph considers a distributed approach to the organisation of key process control elements, and unlike more 'conventional' approaches, seeks to make a fundamental architectural change to the way in which control is organised. We propose this because of the lack of flexibility available in existing, hierarchically-based industrial control systems and hence an inability to easily support operations with changing conditions. Instead, we examine an alternative distributed, non-hierarchical approach to specifying industrial control systems. The specific distributed approach considered in this work – so-called holonic manufacturing approach – has been widely developed in the discrete manufacturing domain as a means of addressing limitations of existing hierarchical control systems. However, the approach has not yet been considered in the process industry domain. The present work thus provides an introduction to the way in which distributed control architectures might be deployed in a process environment.

This monograph presents a distributed approach to the control of process operations that require a high degree of reconfigurability. The key to the approach is to consider a process as comprising a set of readily integrable, modular elements – each of which is able to operate relatively independently and, in control terms, is supported by a degree of stand-alone decision-making capability.

The rationale for the developments reported in this book is that increasing industrial demand for product customisation requires in turn that process operations and hence their control systems should be highly flexible and reconfigurable. This work builds on related recent developments in process control – such as the ISA S88 standard – which seek to enable greater interoperability in batch process control. Such developments also partly address the requirement for increased reconfigurability, yet these developments remain limited in this sense due to the underlying (hierarchical) architecture of today's process control systems. Hierarchical architectures for control systems are known to work well when conditions are orderly, planned and stable but are less effective under rapidly changing situations. For example, frequent arrivals of new production orders, each requiring numerous adjustments to be made at different layers of a control hierarchy, can lead to significant downtime for a process plant whose control systems need to be adjusted at many levels to accommodate such changes. The research in this monograph argues that to effectively address process reconfigurability, a modular, distributed architecture, can provide the necessary means for a process control system to effectively respond to evolving production conditions.

Seeking inspiration from existing distributed approaches to systems management – the so-called Holonic Manufacturing approach in discrete manufacturing and Supply Network Coordination in supply chain management – we present the tools here to enable a distributed process control system capable of reconfigurability. The overall approach is referred to as the Distributed Reconfigurable Process Control or DRPC approach. We make three main contributions in this research which relate to the structure, behaviour and operating strategy of the DRPC system:

i. *A reconfigurable process control architecture:* The architecture comprises four interacting process control elements – called process elements – which are designed to be able to reorganise themselves into alternative processing schemes that can meet the changing production requirements.

ii. *An interaction model for process elements:* The interaction model developed supports plant-wide coordination of process elements and provides these elements with two different approaches for their interaction depending on.

iii. *Distributed process control strategy:* To investigate aspects of the operational behaviour of process elements, an algorithm compatible with the distributed nature of the process elements and their interactions has been developed. By examining a simplified process control problem, it is shown that it is possible to solve typical plant-wide control problems via interactions between distributed process control elements.

Through a systematic evaluation of the proposed approach, we show that the approach presented in this work meets the underpinning business requirements of supporting product/process diversity and enabling dynamic response. In particular, three features of the proposed solution help to achieve this:

a. *Distribution on physical processes:* The reconfigurable control system architecture is developed in a completely distributed form which reflects the physical operations, which then provides increased modularity in process elements.

b. *Separation of structure and behaviour:* All process control algorithms are maintained separate from the underlying control architecture. These algorithms, therefore, can be developed or modified independently of the architecture and vice versa.

c. *Separation of product and equipment:* The product recipe information on 'how to produce a specific product' is retained separate from the equipment control of process units. This recipe information is then integrated dynamically in a distributed manner as and when the production condition change.

This book is presented in three parts. In Part I we introduce the problem being examined and review existing approaches to reconfigurability in a process control setting as well as other related material. In Part II we present the main elements of the DRPC approach and then in Part III we illustrate how the approach might be deployed using an illustrative case example.

The work in this monograph combines an extension to the Ph.D. work of the first author and many years of experience in distributed control research of the second author. The authors would like to acknowledge the support of the Nehru Foundation, Cambridge Commonwealth Trust, the EU IPROMS Network and the UK EPSRC IMRC. We also acknowledge various people involved in this research, including Professor Nick Karcanias, Dr Julian Allwood, Dr Andrew Ogden-Swift, Amro Farid, Professor Vlad Marik, Dr Paul Valckenaars, Wuttiphat Covanich, Alan Thorne, Dr Jin-Lung Chirn and Dr Stefan Bussmann for helpful discussions and suggestions during the course of this work. Our acknowledgement also goes to Neil Macintosh for carefully proofreading the key material, and to Professor Duc Pham at Cardiff, and Anthony Doyle and Simon Rees at Springer-Verlag for providing opportunity to publish this manuscript.

Cambridge, UK
July 2007

Nirav N. Chokshi
Duncan C. McFarlane

Contents

Part I

Problem Development

1

Introduction

1.1 Introduction

"We live in an age of change". This is the opening line of many current articles in the popular press and nowhere is this permanent evolution being more keenly felt than in the industrial sector. Existing under the shadow of relentless cost cutting for decades, it is now clear that no level of price reduction is capable of preserving the industrial status quo. The hope for industries – at least in Western countries – lies in variety, differentiation and customisation – all for little or no extra cost. At the heart of successful industrial change management is being able to adapt and reconfigure operations simply and effectively. That is the essence of this book which focuses on the process industries and in particular how control systems of the future can best support the reconfigurability of process operations. The approach being taken here proposes a fundamental architectural change to the way in which control is organised – we examine a distributed, non-hierarchical approach to specifying industrial control systems. We propose this because of the lack of flexibility demonstrated by existing, hierarchically-based industrial control systems and their subsequent inability to easily support the operations required to adapt to changing conditions.

This book is divided into three parts. In this, the first part, we provide background to the issue of reconfigurability in process control and review the rationale for examining the problem and for choosing to take a distributed approach. We also review existing academic and industrial developments which impact on this area. The second part of the book then presents the main results of this work, introducing the overall approach to distributed, reconfigurable process control (DRPC) and then developing the underlying architecture, interaction mechanisms and control strategies that describe its operations. The final third part of the book assesses the effectiveness of the DRPC development through an illustrative case study.

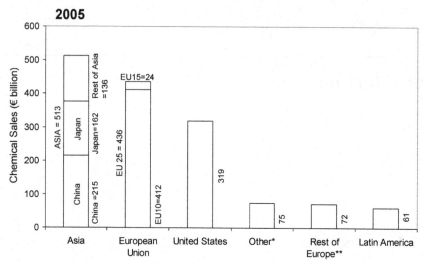

Fig. 1.1. World-wide sales of chemicals (excluding pharmaceuticals)

1.2 Need for Reconfigurable Process Control

To motivate the DRPC development, we begin with an examination of the process industries, the emergence of the need for reconfigurable process control and the current shortcomings of the industry in addressing this need.

1.2.1 Emerging Business Drivers

The process industries form an important industrial sector contributing to the growth of major national economies, both in developed and developing world. Among others, the process sector covers a large spectrum of manufacturing processes including petroleum, petrochemical, pharmaceutical, pulp and paper, consumer goods, metal, utilities and others. Fig. 1.1 shows the scale of world-wide sales of chemicals excluding pharmaceuticals in the year 2005. Within EU the process industry enjoys a strong position with chemicals sector contributing roughly $2 - 2.5\%$ of the GDP and 30% of the world-wide sales of chemicals (Cefic 2006).

Historically, the modern process industry emerged during the second industrial revolution in the early twentieth century, alongside other capital intensive industries such as metals, electrical machines, food processing and automobiles. The early process system designs were based on small-scale unit operations operating in an isolated, *ad hoc* manner. Developments such as *plastic for metal replacement* led to innovation in the industry especially after the second world war (Landau & Arora 1999, Chandler 2005). As demands soared for commodities such as polymers and plastics, the industry enjoyed a rapid growth until the mid-70s when the energy crisis and worldwide recession suddenly hit the process markets with unanticipated shocks. Energy, as the basic ingredient of process businesses, was no longer available cheaply or easily. Nor was it easy to maintain a sustainable demand as markets started to saturate. Many new suppliers from countries having oil reserves (such as Korea, China, Mexico) entered in the market and quickly captured a large share due to low costs. Faced with overcapacity and sunk investment, many companies were forced to reduce costs and improve productivity through restructuring and adapting local processes (Edgar 2004, Chandler 2005). Achieving material and energy efficiency thus became important and started to dominate the agenda of most designers and operators of process businesses and such has remained the case for a period since then.

The last decade has seen a major shift in the way many manufacturing businesses have been managed and process industries have remained no exception to them. Recent surveys conducted in global process industry (Anderson 1997, Bolton & Perris 1999, Backx, Bosgra & Marquardt 2000, Felcht 2002, Shah 2004, Shah 2005) suggest the growing concerns of low-cost customisation and global competition that create new business pressures in consumer sectors such as fast-moving goods and pharmaceuticals where the demand to meet customer expectations has become increasingly significant. Other sectors such as polymers, plastics and petrochemicals are following suit.

It has been claimed (Backx *et al.* 2000, Shah 2005) that to survive against these changing trends the industry will need to move away from conventional mass production type operations to more agile and dynamic processes operating close to the market. Undoubtedly, the emphasis on profitability will continue to hold, yet the driving force in future will shift towards building and maintaining close relations with customers and suppliers in global supply chains. Dynamic reconfiguration of plants, achieved through flexible process designs, will become essential to quickly and efficiently meet the changing demands. As will be important the ability of the plants to produce a variety of product types at time-varying capacities with potentially diverse sources of raw-materials and utilities (Shah 2005).

Influenced by these changing trends, designers and integrators in process engineering have been seeking new and often radical ways of operating process plants. In the fast-moving goods sector, for instance, the trend has been to move away from large, steady-state designs to more discontinuous and dynamic production routines that can be quickly reconfigured; the reuse of equipment through batch and semicontinuous designs has become norm in these sectors and is only likely to continue in future (Keller & Bryan 2000).

Yet, the industry has a long way to progress. To deal with the emerging demands, it will be essential that the plants are made more flexible and reconfigurable. Further support will be necessary to support the dynamic, fast and smooth reorganisation of processes to adapt with the changing market

Table 1.1. Business drivers of future process systems

- MARGIN COMPRESSION AND INTENSIFIED COMPETITION will continue to demand:
    - Competitive production yields
    - Economic consumption of materials, energy and utilities
    - Scalable, resilient on-stream parameters
    - Reduced capital and operational investment
- INCREASINGLY VOLATILE MARKETS will demand:
    - Diverse product portfolio from seasoned to non-standard products
    - Dynamic production and supply networks operating as *extended enterprises*
    - Fast, smooth change management with make-to-order production
    - Responsiveness to disturbances and external/internal variations

---

conditions. The surveys cited above on the changing structure of process industry clearly indicate this shift and suggest a range of parameters where further improvements will be sought. Table 1.1, in summary, gives a short list of key such parameters (adapted from Bolton & Perris 1999). The emphasis on efficient and cost-effective operations will continue, yet the future will ask for increased diversity in products and processes and responsiveness to change and disturbances and to achieve these, a high level of reconfigurability in process designs and control systems. We return to the needs for reconfigurable process control again in more detail in Section 2.3.3.

1.2.2 Shortcomings of Current Industrial Practice

Modern industrial practices in process engineering and control have evolved alongside the changing structure of the industry. The drive for reduced production costs in the late seventies led to increased use of integration and recycles and the need for long spells of steady-state operations. As the sizes of processes increased they also became complex in their design and operations (Lenhoff & Morari 1982). It soon became impossible to control such plants using centralised management structures operating in isolated, silo mode. The concept of hierarchical control was thus born. Multiple levels of control with higher-levels governing the global aspects of production provided the necessary stability and visibility in the business levels that the managers craved for, and so has remained the practice since then.

However, as the industry enters into a new era of customisation and globalisation, the hierarchically-based control structures require a re-assessment. A hierarchical system, by design, works well when conditions internal or external to the production system are orderly, planned and stable but not otherwise, *i.e.,* when disturbances or changes arise. Delays in interpreting information passing between levels in hierarchy result in a control structure that is unable to adapt with the changing conditions where perhaps a sub-optimal but dynamic response may be preferred (Backx *et al.* 2000, McFarlane 1995).

This same argument also extends to engineering and design methods for process engineering. Conventional methods for building physical processes or control systems operate top-down, *i.e.,* they start with the scoping of end-user requirements and building from that a conceptual design that forms the basis of further developments (Douglas 1988). However, it is false to assume that the end-user requirements of today will remain the same tomorrow. The errors and omissions made in the conceptual design thus prevail and as the design progresses, become difficult to rectify. These restrict the later stages in design life cycle to have any influence on the design performance. Process systems built as a basis of only today's requirements thus become sensitive and fail to change, where again perhaps a bottom-up design approach is preferred.

1.2.3 Future Requirements for Process Control

Developing plants of the future that meet the emerging business demands will pose new challenges to practitioners. Achieving fast and smooth changeover between products will require control systems to be made easily reconfigurable, both in their design and in operations. Ability of process systems to react to changing conditions will require the decisions on how to respond to such changes to be distributed down to locations where change occur and not at the higher levels where the visibility to disturbances remains poor and is subjected to delays. As noted by Bolton & Perris (1999), the current approach of learning and improving incrementally from past experiences can keep the industry competitive in short-term, but to sustain and survive in the long-term the industry will need to employ a policy of *strategic learning, i.e.,* doing things not just better, but differently; a fundamental, if not radical, change in the design or operations of process plants will be necessary.

1.3 Motivation for Research

This research is motivated by the need for strategic innovation in the design of process operations in general and process control systems in particular in order to enable simple and feasible reconfiguration. We consider here a distributed approach to developing process control systems in which we construct the control system by combining sets of readily-integrable, modular, autonomous control elements. Each of these control elements – which align with physical functions in the process plant – can operate relatively independent, and in control terms, can support a degree of stand-alone decision-making. The rationale for such a distributed approach is that by constructing the overall system from self-contained modules aligned with process operations, if a process reconfiguration is required, the control elements should be able to be reorganised simply and effectively.

Our investigation was stimulated by similar research in other domains, including discrete manufacturing control, supply chain management, coordination of large-scale power systems and the control of communication networks. These other domains face similar challenges of emerging demands requiring dynamic response to changes both within and external to the system. In particular in discrete manufacturing control the (distributed) concepts of so-called *holonic manufacturing systems* or *agent-based manufacturing control* have received a wider attention in recent years (Christensen 1994, Seidel 1994, van Brussel, Bongaerts, Wyns, Valckenaers & Ginderachter 1999, McFarlane & Bussmann 2000). Recent studies (Mařik & McFarlane 2005, Pĕchouček & Mařik 2006) on the industrial deployment of these new technologies show promising signs for their uptake in industry primarily because of the apparent benefits compared to conventional centralised or hierarchical approaches. However, apart from the work of a small number of authors (*e.g.*, McFarlane 1995, Chokshi & McFarlane 2002, Seilonen, Appelqvist, Vainio, Halme & Koskinen 2002, Niemand 2003), studies on systematic application of these distributed approaches in the process industry are scarce. The research presented in this text hence attempts to bridge this gap by presenting a systematic framework for the application of a distributed approach to continuous process operations requiring a high degree of reconfigurability.

1.4 Outline of the DRPC Approach

The DRPC approach proposed in this text builds upon two key aspects: (i) the so-called concept of *distributed coordination* motivated by the existing research in holonic and agent-based manufacturing and (ii) an analogy between process plants and so-called *virtual enterprises*. The former is considered the primary approach for this work while the latter is considered as a conceptual strategy to help define the design and operational behaviour of distributed control elements. The outcome of the combination of two aspects is an architectural framework that leads to developing the tools necessary for building a reconfigurable process control system.

1.4.1 The Concept of Distributed Coordination

The concept of distributed coordination – coordination of multiple, distributed entities using direct interactions between them – is central to the DRPC approach. The term coordination, used vividly in the routine life, found some early success within industrial control in the 70's and 80's as a method for distributed problem solving (Mesarovic, Macko & Takahara 1970). However, its use in control has been dormant since then. This lack of interest can be attributed more to the success of hierarchical principles in managing large, complex process operations than to the difficulties faced in putting coordination techniques to work as part of on-line control. However, with the ever increasing pace of computing and communication technologies, the principle of coordination, in particular distributed coordination, is thought to provide a successful alternative to hierarchical structures to address the challenges of reconfigurability (Backx *et al.* 2000).

The DRPC approach, similar to the previous research in holonic and agent fields, takes a view of *minimal coupling* between control components is a key to enhancing reconfigurability. This, on one hand, leads to a *distributed* architecture that comprises modular components with stand-alone decision-making, and on the other hand to a *coordination* mechanism that delays the binding commitments linking these components to their run-time operations where these links are established via considering the latest status of operations on the shop-floor. An example of the latter is a separation of product recipe information (how to make a product) from the procedural control of equipment on the plant. The recipe information is then integrated via dynamic, distributed interactions between control elements responsible for each as and when a product is to be produced. It is envisaged that this dual approach will be better positioned to tackle the frequent, time-varying changes expected to arise in high-variety, low-volume industries (such as plastics, polymer, pharmaceuticals) than a hierarchical structure based on multiple-levels of control.

1.4.2 Viewing Process Plants as Virtual Enterprises

While the existing work in holonic or agent-based research is starting to mature, the results therein do not translate directly to process domain due to the tight, finite, physical couplings between process units. In order to extend the existing work, we hence consider an analogy of process plants as being one form of supply chains, in particular, so-called *virtual enterprises* (Camarinha-Matos, Afsarmanesh & Rabelo 2003). A supply chain, similar to a process plant, also involves the network constraints such as transport routes. The success of how well a chain is operating depends on the sharing of information between companies and the coordination of their localised policies (Tayur, Ganeshan & Magazine 1999). Our interest for choosing the analogy then is to look at how companies in a virtual enterprise – being a co-operative alliance – manage and coordinate their operations via mutual interactions; *i.e.*, how do they form, operate and dissolve the alliance in changing times. An analogous model, repeated now at a lower of a process plant, is taken as a conceptual tool to visualise the design and operations of distributed elements in a DRPC system in a manner much similar to the use of so-called *contracting* principle in previous holonic and agent research.

1.4.3 Research Contributions

In the course of developing the DRPC approach presented here, we make three main contributions to the fields of reconfigurable process control and holonic and agent-based manufacturing.

i. *A reconfigurable process control architecture:* A new control architecture is proposed as comprising four interacting process control elements – called process elements – which are designed to be able to reorganise themselves into alternative processing schemes that can meet the changing production requirements.

ii. *An interaction model for process elements:* To support the plant-wide coordination of process elements, an interaction model is developed that provides these elements with the interaction approach to build and operate alternative process schemes.

iii. *A Distributed process control strategy:* To investigate aspects of the operational behaviour of process elements, an algorithm compatible with the distributed nature of process elements and their interactions is proposed. By examining a simplified process control problem, it is shown that it is possible to solve typical plant-wide control problems via interactions between distributed process control elements.

Combining these three developments together provides an underpinning framework for a more detailed specification of a reconfigurable process control system. In order to motivate how such a system might operate, the next section outlines a hypothetical vision for the process control system of the future.

1.4.4 A Long-term Vision of a DRPC System

To quickly illustrate how a DRPC approach might work, we describe below a long-term, imaginary vision of a reconfigurable process control system built based upon it. The vision is deliberately taken to the extreme to demonstrate the potential range of achievements that can be made.

Plant Design

A reconfigurable process plant built using DRPC approach comprises an assortment of multiple dedicated and flexible process entities (the so-called process elements) integrated together in a flexible configuration. The process elements are modular and integrated in a bottom-up manner. The dedicated elements, such as main reactor unit, transfer equipment *etc.*, provide the scalable production throughput at optimum operating costs. The flexible elements are of 'plug-and-produce' nature and possess the multipurpose functionality to support their re-use in producing a diverse mix of products or product grades. These elements can either be designed *a priori* and included in the process or can be integrated during run-time operations as and when needed. This facility allows replacing a dedicated process element with a combination of multiple flexible elements if necessary.

Control Design

Each process element possesses its own control and coordination modules. Compared to conventional control hierarchy, each level in the hierarchy is split along the physical dimension (*i.e.*, individual process units, process headers *etc.*), followed by vertically integrating the local blocks into the control and coordination modules of the process elements. Fig. 1.2 illustrates this decomposition. The coordination modules are associated with the decision-making levels in the hierarchy (*i.e.*, planning, scheduling and optimisation) while the control modules with the execution levels (*i.e.*, basic control and the interface to the physical process).

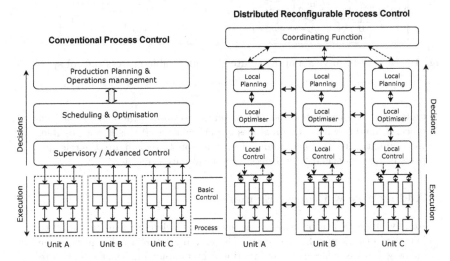

Fig. 1.2. Conventional control vs. Distributed reconfigurable process control

Plant Operations

Fig. 1.3 – in abstract – depicts the nature of operations of process elements in a reconfigurable process plant. When idle, the process elements wait for a production order to arrive. On arrival of an order, a new product element is created representing the order requirements, *e.g.*, quality and throughput requirement, product recipe. Multiple product elements can co-exist, however only a few will be produced at a time. Each product element together with other process elements representing the available production capabilities then carry out a round of distributed interactions to obtain a process scheme that can meet the production requirements for that order. The elements may follow an economic production goal to arrive at the choice.

The elements involved subsequently reorganise the physical process as agreed. The actual production then begins when advised by the product element. During operations, if a deviation occurs, *e.g.*, one of the process

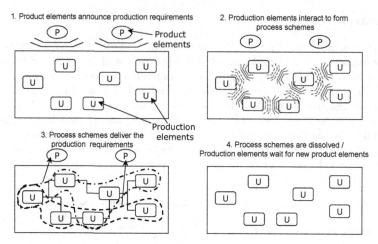

Fig. 1.3. A long-term vision of distributed reconfigurable process control approach

units fails, the associated elements attempt to absorb the disruption in a graceful manner with a minimal loss of performance. If this is not possible, the elements initiate a new round of interactions to reconfigure the relevant parts of process scheme, or if necessary, terminate the order.

Adding a New Process Element

At any stage in the operations, a new process element can be added to join the network of other elements. The arriving element takes part in the ongoing process schemes (or interactions) by announcing its capabilities to other process elements. The elements that can make use of its capabilities then interact to identify a switch to an alternative configuration of the relevant process schemes.

Terminate Production Order

Once the order requirements are met, the process element engaged in the order dissolve the process scheme. Those elements involved in other schemes reconfigure their operations as necessary.

As generally evident from the above description, in a DRPC approach the plantwide response to normal or abnormal situations emerge from localised actions of process elements acting within dynamically organised groups as part of process schemes. This is conceptually different from conventional hierarchical models where the response remains governed by higher-levels plans and schedules that transcend top-down; a change in plant condition may require complete reschedule or even replan (if a minor change or disturbance occurs) or taking plants offline (if a major system fails).

1.5 Structure of the Monograph

As discussed earlier, this monograph is structured in three parts: (i) problem development, (ii) a distributed reconfigurable process control approach and (iii) an assessment of the DRPC approach. The three parts comprise a total of eight chapters. Fig. 1.4 outlines the structure.

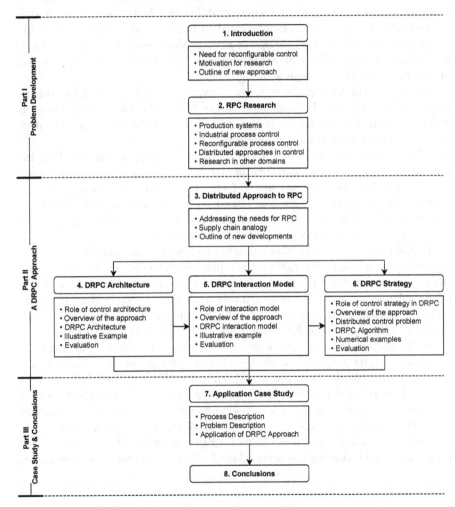

Fig. 1.4. Structure of the monograph

Chapter 2 starts the discussion by assessing the existing structure of industrial process control systems. Reconfigurable process control, the topic of this monograph, is then examined in order to understand the needs for reconfigurability and the requirements they place on the design of process systems, in particular process control systems. Numerous solution approaches developed in the past in distributed control research are next reviewed, with an extended focus on the holonic and multi-agent systems research. Finally, the experiences learnt from other industrial or non-industrial domains are evaluated to motivate the present work and to seek inspiration from the solution concepts they can offer to improve reconfigurability of control systems.

Chapter 3 next provides an outline of the key developments in Part II. It first reassesses the needs for reconfigurability and links them with the distributed coordination concepts suggested in holonic and related other fields in order to identify the exact scope of work in this research and to provide a conceptual overview of a how a reconfigurable process control system based on the proposed DRPC approach might operate.

Chapter 4, the first in the series of three chapters, presents a control architecture for defining the basic elements in a reconfigurable process system and their control functionality. It also defines the structure of process control system resulting from their combination.

Chapter 5 next presents an interaction model to support the real-time interactions of process elements. The model defines the information exchange mechanisms that the elements use to form and operate (temporary) process schemes. A particular emphasis is placed on the manner in which these elements identify a feasible process scheme from the details given in product recipes and the production capabilities available in the plant.

Chapter 6 associates the interaction model from the previous chapters with a distributed solution strategy that the process elements can use to identify their local operating settings. To formalise the strategy mathematically, a simplified process control problem involving a linear, steady-state dynamic model of process units is considered. The concept of so-called nested decomposition is then borrowed from the optimisation literature to systematically define the strategy in the form of an iterative algorithm. Numerical examples demonstrating its operation in small-scale examples are presented based on a software prototype developed in MATLAB®. Results of this analysis are also validated against the benchmark of equivalent centralised implementation.

Chapter 7 brings together the developments in the previous three chapters by applying them to an industrial case study of a multipurpose process plant.

Chapter 8 finally concludes the monograph by summarising the key contributions and identifying the areas where further developments can be made.

2

Reconfigurable Process Control Research

2.1 Introduction

In this chapter we collect together a number of different developments which lay the foundations for the distributed, reconfigurable process control (DRPC) approach we are proposing. We begin by positioning process operations within the spectrum of industrial production approaches and in particular provide a contrast between continuous process and discrete manufacturing. (This is important when reviewing existing work in distributed reconfigurable control.) We then examine the evolution of process control and in particular developments which have dealt with reconfigurability challenges and their limitations. The second part of the chapter then goes on to introduce distributed coordination methods in process control and then to provide a comprehensive review of the way in which distributed coordination has been applied in other industrial domains.

2.2 Classification of Manufacturing Systems

Manufacturing industries involve a range of production operations and operating conditions. Based on the physical layout of production processes these can be split broadly into: discrete parts manufacturing (automobile, semiconductor industries) and continuous processes (polymer, pharmaceuticals, petroleum industries).

In a discrete process, the individual parts are produced first using various discrete, loosely coupled operations such as machining, drilling, grinding *etc.* These parts are then pieced together in an assembly line to create the main end-product. Often a large number of parts may be involved (*e.g.*, in a car engine) with parts, being physically stable in nature, can be stored or transferred between lines. Unfinished orders can be pre-empted or transferred for more important orders where facility exists.

Table 2.1. Production control in discrete and continuous processes

	DISCRETE	CONTINUOUS
Physical Layout	Jobshop/flowshop with parallel machines	Line / series of equipment
Objective	Part or job centered	Product or recipe centered
Coupling	Intermediate buffers due to conveyors,AGVs	Tightly coupled with piping network
	Time/schedule based	Product based (non-mixing)
	Stable intermediate forms of parts	Possibly unstable chemistry
Controlled Variables	Due date, arrival time, processing time	Process values / set-points, product quality
Control Freedom	Machine assignment, route flexibility	Equipment operational modes, route Flexibility
Control Strategy	Discrete on-off logic (using PLC)	PID/multivariable control (using DCS/PLC)
Example	Semiconductor, Automobile	Petrochemicals, Polymer

A continuous process instead involves continuous flow of materials (such as bulk chemicals) and utilities through process units interconnected via piping streams. New property values are added to these streams as they pass through process units. Normally, an interim form of the end-product is first produced using one or more reaction operations. The un-reacted raw-materials are then separated and re-used while the interim product is purified and processed to bring into final form. The interim product can be mostly unstable and may not sustain long storage. Therefore, pre-empting or transferring of unfinished orders is not normally possible.

These physical differences between discrete and continuous processes lead to their use of different production goals and control methods as summarised in Table 2.1. In a discrete process, the target is to identify a routing of discrete parts across shop-floor and assign appropriate tasks to machines and define their scheduling. In a continuous process the routing remains normally fixed, and the goal instead is to identify the local operating settings of process units and their combinations across the plant that meet the required quality and throughput of the end-product.

A misconception generally prevails, particularly in the research community, that continuous processes are primarily long-term, steady-state operations. This is strictly not true however. By shortening the range and horizon of operations, a continuous process can be made to behave as discontinuous or discrete as in a batch process. Fig. 2.1 depicts the spectrum of discontinuous operations that can be found in process industries. As Keller & Bryan (2000) note, almost half of the production tonnage in process industry comes from discontinuous processes – the proportion which is only likely to grow in future.

Particularly important to this monograph is the so-called *semicontinuous class* of processes in Fig. 2.1, which similar to a continuous process also involve continuous flow of materials and utilities, however the plants are not operated

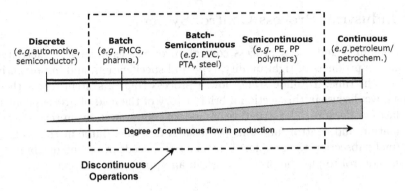

Fig. 2.1. Discontinuous operations in process industry

in a purely steady-state mode. Instead, so-called *campaign* mode of operation is often used (Papageorgiou & Pantelides 1996). The overall planning horizon in a campaign operation is split into multiple product campaigns, each associated with a different product, product grade and/or raw-materials. Subsequently, a campaign for any one product is first produced for a defined period. The production conditions are then changed and a separate campaign for another product is produced using the same set of equipment. Thus, although each campaign operates in a continuous mode, the sequence of campaigns over a certain period results in discontinuity of operations.

The key rationale for the move towards discontinuous operations has been to increase the re-use of equipment particularly when a number of products are to be produced in typically small amounts that do not justify the use of stand-alone plants. The level of re-use required and therefore the plant design may vary depending on the number and type of products to be produced and the variations expected in market demands. In a so-called *multiproduct* design the process is organised such that all products follow the same path and use the same equipment with typically one product produced at a time. In a more flexible *multipurpose* design each product may take one or more distinct processing paths with possibly more than one products produced together where necessary. Multiproduct designs are thus suitable for conditions when the products and processes to be used are known in advance but the quantities or time scales are not, whereas multipurpose designs are suitable when none of these is known and the plant is constructed to contain equipment suitable for certain unit operations with a range of parameters (Mah 1990).

The management of (dis)continuous operations involves a great deal more control operations than purely continuous processes in order to define which products to be produced, when and how. To understand this role of control more clearly, we next discuss the evolution in process control and the structure of modern process control systems.

2.3 Industrial Process Control Systems

The domain of industrial process control encompasses a range of activities to produce products of right quality, type and specification, and importantly, at the right time. To understand how a process control system meets these targets, we discuss in this section a brief history of the field of process control over last few decades. The structure of modern control systems in terms of the information and control functions involved is described later in the section. The final subsection then explores the emerging needs of reconfigurability in process control to put the present work in an appropriate context.

2.3.1 Evolution in Industrial Process Control

Modern process control systems came into existence after various phases of evolution, with each phase having a distinct impression on a particular aspect of the system design. The early designs were governed by then-current business drivers, or indeed the inhibitors such as energy crisis, but in recent years numerous other factors such as IT and communication technologies have played their role in changing the perception of process control in the industry.

The early control systems developed in the 1950's or before were focussed on regulatory control, *i.e.,* the PID controller was the key building block of control. The advent of mechanical and pneumatic devices at this time and subsequently electronic controllers in the early 60's allowed a level of remote control to be achieved but the scope of control was limited to a single or at most few process variables. Coordination of unit systems was mainly the operator's responsibility.

The first major shift in process control occurred in the 1960's when digital circuits and computers were introduced. In a so-called *Direct Digital Control* (DDC) application, a computer was used to replace analog controllers and panelboard displays. This was a pioneering change in control as it allowed advanced strategies, such as sampled data control, to be employed as part regulatory functions. However, the centralised role of computer was a risk of failure with possible complete loss of control. The costs and skills required to deploy computers were also prohibitive (around $30,000-250,000) and proved difficult to justify against simple, PID control in most cases (Smith 1970).

By the early 1970's it became clear that putting computers (or the optimal controllers developed based upon them) in direct control of physical processes is neither convenient nor necessary when a two-level scheme is employed comprising a supervisory (optimising) controller and the bottom-level regulatory controllers. The supervisory computer would focus on key variables (such as reactor yield) and provide regulatory controllers with set points for implementation through analog or digital hardware. Since the computer did not replace underlying hardware, its failure was not critical (Edgar 2004). The use of supervisory control was a conceptual change in the design of control

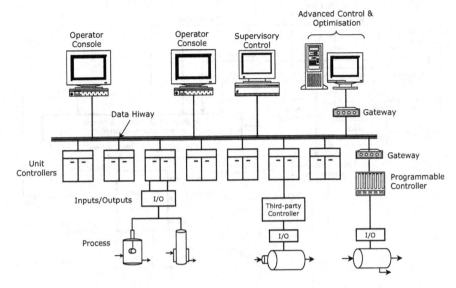

Fig. 2.2. Architecture of a distributed control system (DCS)

system architectures that led to the birth of so-called *distributed control systems* as shown in Fig. 2.2. The significant reduction in size and costs achieved by DCSs and the parallel development of so-called *programmable logic controllers* (PLCs) led to the widespread acceptance of these new architectures in the industry (Samad, McLaughlin & Lu 2007).

A number of events occurred in the 1970's and 80's that changed the perception and hence the structure of process control in the industry. The energy crisis in the mid-70s had a profound influence in this as energy was no longer available cheaply or easily, nor was it easy to sustain long-term demand as new suppliers entered in the markets from countries having access to oil reserves. Sunk with overcapacity and costs many enterprises, particularly in western world, were forced to restructure their businesses. While the economies of scale and scope still remained the dominant means for cutting costs, it became clear that further reduction could only be achieved by reducing the material and energy consumptions and importantly, from making improvements in process control. New control functions such as planning and scheduling, statistical process control, optimisation *etc.*, thus started to take shape as part of the mainstream components of production control and have remained so for the time since then (Chandler 2005).

The period of 1990's saw process control moving one step further from that of managing individual plants to managing enterprise-wide functions. The integration of online enterprise data consisting commercial and financial information with the real-time functions of planning, scheduling and control has become significant (Shobrys & White 2002). Equally significant has been

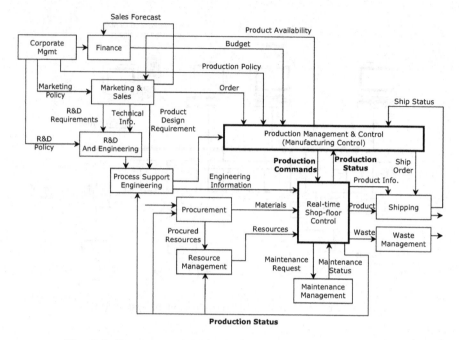

Fig. 2.3. Process control interface to other enterprise functions

the need to create *open* system architectures that enable the system integrators to mix-and-match control components from different suppliers and technologies in a seamless fashion. The adoption of open technologies such as OLE for process control, fieldbus networks and Commercial-off-the-Shelf tools (*e.g.*, based on Microsoft Windows platform) have become the norm in many cases for developing modular system designs that can be rapidly engineered and reconfigured (O'Brien & Woll 2005).

In summary, the past six decades of history have seen process control grow from a primitive regulatory mechanism to a function central to an enterprise that provides the means necessary to deliver the emerging business goals in changing times.

2.3.2 Key Features of Modern Process Control Systems

Today's state-of-the-art process control system includes a variety of tools and techniques to control plant(s) comprising multiple, interconnected unit operations. The control system also interfaces with numerous other enterprise functions shown in Fig. 2.3. In this section, we now discuss the key structural aspects of the modern systems to examine the underlying information flows in coordinating the plantwide operations.

Structure of Process Control Hierarchy

The structure of modern process control systems is based on a hierarchical approach developed as part of wider Computer Integrated Manufacturing (CIM) initiatives in the early 80's. A system hierarchy was preferred as a suitable, and at times a necessary, mechanism to deal with the growing complexity and size of process systems that involved control problems spanning hundreds of variables.

The design of a hierarchical control system has been structured around a *functional* hierarchy that decomposes the business goals defining which products to produce, when and how, to lowest-level set points for regulatory control. Multiple levels of decomposition may be used with each level fixing certain key variables. The implementation of the resulting goals is then carried out via an *aggregation* hierarchy that, in most cases, parallels to the physical decomposition of a plant into its constituent elements (*i.e.*, area, cell, units *etc.*). Fig. 2.4 depicts the two forms of hierarchy using a so-called *Purdue Reference Model* (PRM) employed widely in the industry (Williams 1989).

In both forms of hierarchy, the controllers at successively higher levels cover the larger and broader but relatively slower aspects of overall system behaviour to provide the visibility to global, long-term operations. The decision-time horizon of higher levels also remain longer than those of lower levels. To limit the size of problem formulations, the higher level problem descriptions are generally less structured, involving more uncertainties, than those for lower levels (Mesarovic *et al.* 1970).

The control of production in a hierarchical system under both normal and abnormal conditions is governed by hierarchical communication. When situations are normal, the business goals are propagated to lower levels where decisions at each level are made based on the fixed parameters. When an error occurs, the controller responsible for that level attempts to resolve the uncertainty. If this is not achievable, the higher level controller is invoked to alter the decisions on these fixed parameters. If in turn, the error can still be not resolved, the problem passes up a further level and so on. Hierarchical control thus provides a level of visibility in production operations when conditions are planned and stable, and a level of flexibility in decisions when contingencies arise. Historically, these attributes have underpinned the success of hierarchical control in mass production environment where operations remain largely steady-state and the focus of production control is to economise the production costs through planned, stable, long-term operations.

Information Flow and Coordination

Looking further in detail at Fig. 2.4, the control functions in a hierarchical system can be split into two categories: levels 4 and 3, referred to collectively as *manufacturing control*, and levels 2, 1 and 0, referred to as *real-time*

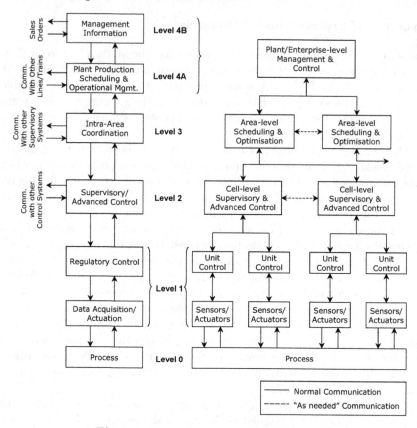

Fig. 2.4. Hierarchical control architecture

control. The manufacturing control levels are responsible for decisions on production management (*i.e.*, which products to produce and when) and the coordination of product flows (*i.e.*, how to produce these products), while the real-time control levels are responsible for executing the outcomes of these decisions onto physical process. These functions also feedback the necessary plant information back to higher levels.

The research in this monograph is mainly focussed on manufacturing control levels, *i.e.*, levels 3 and 4 and the interface between the two. Fig. 2.5 shows the different categories of information involved at these levels, which can be described as follows (adapted from ANSI/ISA 2003):

- *Product Definition Information:* This covers the information on product production rules and the bill of materials and resources required in production. The production rules are abstract and only define how a chemist would produce the product on a laboratory scale, *i.e.*, the information about materials involved in unit operations and their operating conditions.

Fig. 2.5. Areas of information exchange at levels 3 and 4

- *Production Capability Information :* This represents the capability information for production resources available on the plant, such as equipment, materials, personnel, energy, consumables *etc.* The information may include details about their design and operational attributes, their current maintenance status and the capacity scheduled for near future.
- *Production Information:* This defines the information necessary to facilitate the actual production. It may cover the areas such as production history, in-process inventory, scheduling of equipment, and the detailed operating procedures *etc.*

The product definition information (how to make a product) is interpreted in terms of the production capability information (what is available) to define the production information (what to make and results). In traditional or legacy systems this integration may be carried out by an operator or a planning and scheduling function operating offline and using spreadsheet tools and/or human knowledge. In more modern systems the standards such as ISA-S88 (ANSI/ISA 1995) and ISA-S95 (ANSI/ISA 2003) are employed to speed up the process and support rapid integration. The key to rapid integration in both standards is the separation of production rules (so-called *recipes* in ISA-S88) from production capabilities of equipment and other resources in the plant. The separation enables creating site-independent, generic recipes that can be deployed across different sites, situated perhaps in different countries and/or having access to different types of resources. As discussed later in Section 2.4, this principle of separating product (recipe) information from that of the processing operations has also been identified as being key for enhancing reconfigurability elsewhere – for example in the holonic and agent-based industrial control research (van Brussel, Wyns, Valckenaers, Bongaerts

& Peeters 1998, Chirn & McFarlane 2001). The separation, in turn, also forms one of the key principles in developing the DRPC approach.

2.3.3 Reconfigurability in Process Control

We now revisit the main topic of this monograph – reconfigurable process control – to understand the incentives for enhancing the reconfigurability of process operations and the factors that characterise reconfigurability in terms of underpinning system requirements.

In a dictionary sense, the term *reconfigurability* of a (computing) system can be defined as its ability to adapt to a new task by altering its configuration (based on Oxford English Dictionary (2005) definition of *to reconfigure*). In the context of production control, reconfigurability then refers to the ability of the control system to adapt to emerging changes (*e.g.*, introduction of new products, processes, raw-materials, utilities, technologies) or disturbances in production operations (*e.g.*, changes in market demands, prices, failure of a process unit, loss or raw-material or utility supplies).

The term *reconfigurable process control* (in short RPC) defines a paradigm in the design of process control systems where reconfigurability forms an essential criteria of the design process. Intuitively, it translates to a facility in the design method with which the control elements can be (i) decoupled, (ii) reorganised, and (iii) recoupled into a new configuration in a possibly smooth and transient manner. The type and nature of reconfigurability required may depend on the ultimate needs of the specific application and the trade-offs that it may have with other design goals. An RPC approach for control design thus provides a layer of additional design decisions that combined with other design criteria and fundamental technical principles should lead to a required level of reconfigurability in the design of control operations.

To develop a new RPC approach, we must therefore understand the motives for introducing reconfigurability in process control. In broad terms, these can be divided into three categories: (i) business needs, (ii) engineer and design needs and (iii) operational needs from end-users.

Business Needs for Reconfigurability

The business needs for reconfigurability emerge from the changing structure of global process industry, *i.e.*, the increased attention on product customisation and globalisation in recent years with a move towards service-centric operations.

As generally true, the process industry sits in the middle of wider supply chains (such as in semiconductor, automotive, consumer goods *etc.*) and faces the impact of technological growth, not just within its own, but also in other industries. With the emergence of new manufacturing technologies and increased pace of technological change (*e.g.*, in electronics industry), the demand patterns of consumers of process industries have been constantly changing. For example, the inventions in mobile phones, computing, audio/visual equipment, home appliances and consumer goods, *etc.*, all nowadays require new varieties of basic products (*i.e.*, polymers, plastic) with additional features, high product quality and better service life. Against this increased variety, the demand for conventional products and commodities has also been sustained or even increased over the past few years as a result of the growing demand from emerging economies in the developing world (Cefic 2006).

However, as Shah (2005) rightly points out, production systems or supply chains in process businesses have yet to catch up with these changing trends or the widening scope of operations. Performance benchmarks for process supply chains generally do not compare well with other sectors (*e.g.*, automotive), for example:

- the stock levels in the chain amount to $30 - 90\%$ of annual demand, with usually $4 - 24$ weeks' worth of finished good stocks in 'pipeline';
- the supply chain cycle times (time elapsed between raw-materials entering and products leaving the chain) tend to lie between 1000 to 8000 hours, of which only 0.3-5% actually involve value-adding operations;
- the material efficiencies tend to be low or below average with only a small proportion of materials entering the chain end up as final products (in case of fine chemicals and pharmaceuticals this figure can be as low as 1-10%).

Clearly, there are incentives to improve here, but large improvements cannot be achieved simply by changing the logistics or transactional processes in supply chains. Rather, some fundamental changes are necessary, particularly at the process and plant level and at the interfaces between various constituents of the value chains (Shah 2005). To a manager responsible for a process enterprise, this means some new challenges for reconfigurable operations:

- shorter product life cycles, with shorter time from innovate-to-market;
- diverse product portfolio with a drive to deliver specialty products at commodity prices;
- enhanced relations with suppliers and customers in global supply chains.

Engineering and Design Needs for Reconfigurability

Even if the business demands of today have been the same as they were twenty years ago, still there are reasons for building reconfigurable process designs from an engineering and design point of view, especially with having the benefits of all technical knowhow gained over the years.

As stated earlier in Chapter 1, the peril of conventional design techniques, both in process systems engineering and control (see, for example, Douglas

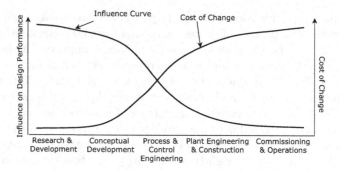

Fig. 2.6. Cost of change across process life cycle

1988, Biegler, Grossmann & Westerberg 1997), comes from their use of a top-down method for scoping the end-user requirements and building from that a conceptual design that forms the basis of further developments. While this approach certainly aids in visibility to the subsequent design phases, the process or control designs built as a basis of conceptual design can become customised and susceptible to change as the design progresses as shown by the 'cost of change' curve in Fig. 2.6. Instead, a combined approach of top-down decomposition of requirements followed by bottom-up integration of standardised components would be preferred as it can support the design modifications at any stage in the life-cycle.

The use of standardised, reusable designs is also preferred to more customised or bespoke designs by the developers of process and control components (Schug & Realff 1996). While customised designs match the requirements of specific applications and incur sale (*e.g.*, in replacing an existing kit), they also need repeating the same design effort and regulatory approval time and expenses. This can be cumbersome in safety or quality critical applications (such as in nuclear, chemicals and pharmaceuticals industries) where standardised designs may be preferred as they can be re-used with shorter lead times and lower engineering costs.

With the increasing pace of technological advances, there also remains a scope for introducing new technologies, *e.g.*, IT and communications, to avoid obsolescence. Often, the new technologies are also more efficient, cheap to procure and easy to build. However, the benchmarks in this case also do not compare well against, for example, to those in automotive or semiconductor industry. In producing chemicals and plastics, the capital and raw-materials cost as much as $50 - 60\%$. Because the plants cost so much, they are usually run for many years and only upgraded when obsolete. This often means lost opportunities. Instead, many lessons can be learnt from experiences in the automobile industry where the use of cheap sensors and on-board computers has transformed motor cars into more comfortable and reliable machines that are also economical to build (Anderson 1997).

In summary, the engineering and design needs for reconfigurability are:

- support for design modifications, during and after the design life-cycle
- use of standardised, re-usable designs with shorter lead-times
- support for technological advances

Operational Needs for Reconfigurability

With increased emphasis on material and energy conservation, it has been a common practice in recent years to design plants with reduced losses, *i.e.*, the use of recycles and heat integration has been norm for a while. While such measures do work in practice and deliver the end results of reduced investment and inventory, they also add up to the operational costs because fluids need to be pumped around constantly. More importantly, they lead to stronger interactions between process units that often cause operational difficulties particularly during transients (Lenhoff & Morari 1982). To maintain satisfactory performance, the plants are hence designed with tighter margins and run in steady-state modes for longer periods. In practical situations, with increased emphasis on product and process variety, the design efficiency can however be a secondary concern. The primary concern instead is to make processes flexible, operable and controllable to handle product/process changeovers or internal and external disturbances such as changes in demands, market prices or arrival of new opportunities (Shah 2005). Many of these require invariably some changes in conventional practices.

On reliability of operations, it has also been a practice to assume that process components are unreliable and that operational upsets are likely to occur, hence redundancy is considered by default (Koolen 1998). Although this helps keep the plants running unattended, it also means the inclusion of spare equipment, devices and sensors. More often this can be avoided if equipment functions are simplified and combined into multipurpose equipment (such as reactive distillation) or broken down into manageable, modular functions that can enhance transparency of operations without compromising on reliability (Schug & Realff 1996).

But, as with any other system, failures do occur, *e.g.*, a process unit fails or becomes bottleneck due to its age or frequent use. Whilst plants or control systems built with redundancy can tackle failures better, there always remains a scope for a level of built-in *fault-tolerance*, *i.e.*, the ability to provide graceful degradation of performance, and where possible, support easy recovery or replacement of failed component. This also is a reconfigurability issue as the losses from a failure or recovering from a failure can sometimes outweigh the cost of the equipment or control system itself.

To summarise, the reconfigurability needs from the perspective of an end-user responsible for operating a process plant are:

- transparent design that is easy to comprehend and operate
- flexible, operable design that supports easy changeover management and disturbance handling

		Diversity	Modifiability	Responsiveness	Fault-tolerance
Business Needs	• Shorter product life-cycles	✓	✓	✓	
	• Product customisation & differentiation	✓	✓		
	• Enhanced supply chain relations	✓		✓	
Engineering & Design Needs	• Ease of design modifications		✓		
	• Standardised, re-usable designs		✓		✓
	• Support for technological advances	✓	✓		✓
Operational Needs	• Transparent operations		✓	✓	✓
	• Support for changeover/disturbance handling	✓	✓	✓	
	• Graceful degradation of performance during failures		✓	✓	✓

Fig. 2.7. System requirements for reconfigurable process control (the shaded labels show a major link for all four system properties)

- fault-tolerant design with graceful degradation of performance when failure occurs

Summary

Focussing particularly on process control, the reconfigurability needs identified in this section can be summarised into four key system properties as shown in Fig. 2.7 and defined below:

- **Diversity**: The ability to introduce new products and processes including raw-materials, utilities and product recipes;
- **Modifiability**: The ability to support ready integration of new components or the reorganisation of existing components;
- **Responsiveness**: The ability to provide a timely response to product changeovers or disturbances or to adapt to new plant conditions;
- **Fault-tolerance**: The ability to tolerate failures or disturbances and when necessary provide graceful degradation of performance.

While diversity and modifiability are more static properties that concern with the underlying architecture and information flows between control elements, responsiveness and fault-tolerance are both static and dynamic measures and relate to how well the control system is able to cope with dynamic changes, disturbances or failures. We believe a process control system that possesses the above properties should have a high degree of reconfigurability. It is for this reason that we focus this work on distributed coordination methods – which are reviewed next.

2.4 Distributed Approaches in Control

This research presents a distributed approach to reconfigurable process control. In order to understand the rationale for taking such an approach, we now discuss the general concepts behind distributed control approaches developed in the past and in particular, examine in the so-called *holonic manufacturing* and *agent-based* control fields.

2.4.1 Understanding Distributed Control

The concept of *distribution* in control, sometimes referred to as *decentralised* control, is rooted in large-scale and complex systems such as power networks, communication networks, markets and organisations. In such large systems, the standard presupposition for control that *information about the system, or calculations based upon it, are available centrally in a single location* does not often hold. In some cases it may be impossible to collect all information centrally (*e.g.*, in case of markets, the companies may prefer not to disclose their internal details to others) or in other cases the information transfer may have an economic or reliability cost which cannot be ignored (Šiljak 1991). In general though, it remains important that the system is flexible and robust enough to absorb various and sudden changes and be able to accommodate graceful failures in components where a centralised decision system can easily fail (Androulakis & Reklaitis 1999).

A distributed control or decision-making system circumvents this information constraint of a large-scale or complex system by spreading the control calculations or decisions directly to the locations where information exists. The process of distribution generally follows three key principles:

i. *Decomposition:* The overall system is split into multiple subsystems such that variables local to any subsystem are strongly coupled while those among subsystems are only weakly coupled; the term *coupling* here may refer to the impact that a change in any variable has on other variables;

ii. *Local decisions:* Each subsystem is associated with a local decision-making agent or controller that possesses the knowledge of its own subsystem plus at most a partial knowledge of its neighboring subsystems; these local controllers may work towards their individual control objectives or to a team objective or to a combination of both;

iii. *Coordination:* The impact of local actions of any controller on other subsystems is assessed and where necessary, coordinated via some form of communication to solve the local problems or a common, global problem or a combination of both; the communication may be either *direct* (through communication links) or indirect (via observing the perturbations from other subsystems).

Process plants, in one sense, can be perceived as a form of large-scale, complex systems because of their highly interconnected nature. A process control problem, if cast as a computational problem, would exhibit this large-scale behaviour in terms of its model coefficients, *e.g.*, a large number of model elements referring to piping connections between process units would be either small or zero in value. This suggests that a process control problem might be decomposed and solved – in principle – in a similar distributed manner. In modern DCS architectures this assertion has been used – at least partially – to implement the bottom regulatory control level in a distributed form. A similar interest is also growing to distribute the other levels in the hierarchy (see, for example, Camponogara, Jia, Krogh & Talukdar 2002, Lu 2003, Venkat, Hiskens, Rawlings & Wright 2006) and the planning and control problems concerning process supply chains (Perea-López, Grossmann, Ydstie & Tahmassebi 2001).

2.4.2 Solution Techniques for Distributed Control

The solution approaches developed in the past for distributed control – while all follow the above-mentioned three principles – differ in the way the local problems are defined and coordinated across the system. Based on the type of coordination mechanism used for problem solving these can be split broadly into so-called *hierarchical coordination* and *distributed coordination* techniques.

Hierarchical Coordination

In a hierarchical, or so-called *multi-level* scheme, the coordination is achieved by a separate higher level controller. Each local controller receives a freedom to choose its control actions based on its local system model and cost criterion, both derived from a simplification of the overall model and cost criterion. In order that these independently arrived choices are coherent, a separate higher-level controller or so-called *coordinator* is used which incrementally adjusts the individual models or criteria such that the combined cost for the whole system improves. The interactions thus repeat between two levels until a form of convergence is achieved.

Research in hierarchical coordination received wide interest in the 60's and 70's when it was difficult to solve large-scale linear programs using limited computing facilities available then. The first known coordination or so-called *decomposition* algorithm is due to Dantzig & Wolfe (1961) where distribution was used to solve large-scale planning problems via coordination. A *dual* method was suggested therein where the coordinator adjusts Lagrange multipliers or so-called *marginal costs* for coupling constraints associating the local subsystems. Benders (1962) proposed the first *primal* algorithm for linear programs that was later generalised by Geoffrion (Geoffrion 1970, Geoffrion 1972) for wider class of non-linear problems. In a primal scheme the coordinator directly fixes the coupling variables connecting the local subsystems so as to incrementally refine the bounds within which the local controllers can choose their actions. Numerous coordination algorithms and solution techniques have been developed since this early work for applications in operations research and later in systems theory and control engineering. See (Mesarovic *et al.* 1970, Findeisen, Bailey, Brdys, Malinowski, Tatjewski & Wozniak 1980, Jamshidi 1983) for detailed overviews.

Application of hierarchical coordination in process applications has been scattered throughout the years. The early references include (Brosilow & Lasdon 1965, Lasdon 1970, Morari, Arkun & Stephanopoulos 1980). More recently, Katebi & Johnson (1997) considered a dual method for optimising control of chemical processes. Jose & Ungar (2000) applied the so-called *Slack Auction* method to process optimisation where a purpose-built auction mechanism was used to coordinate the interaction variables associated with piping connections between process units. Grothey (2001) proposed a mixed primal-dual technique in a *fixed-and-price* algorithm for more general class of process control problems of nonlinear form. Hou (2001) applied a dual method for coordinating large-scale neural network problems arising in optimal control.

It is worth noting that the above *multi-level* schemes are different than *multi-layer* schemes used in conventional control hierarchy (Fig. 2.4). In a multi-layer scheme, the higher-level controller solves the same plantwide problem, but at an aggregate level, to fix certain key variables. In a multi-level scheme the coordinator is not required to solve any such problem. This has an advantage that modifications required in any part of the system are only made at the local level. The coordinator, being a centralised function, however still poses a threat of single point of failure. Also, the process of coordination is a synchronous process and can be limiting as all local solutions problems must be communicated to coordinator before it can adjust local models or cost criteria. The computational speed of the overall problem can thus be limited by the slowest or busiest local processor among all.

Distributed Coordination

In a distributed coordination scheme, the role of coordinator is removed. Instead the coordination is achieved by the decision-making controllers themselves (called below as *agents*). The agents interact in a distributed mode and are guided by some form of global rule that leads them to converge towards a consensus.

Central to distributed coordination is the information that agents exchange in making local decisions. Agents may not communicate at all and still reach concensus by using some form of min-max strategy of choosing local decisions that satisfy the worst-case physical interactions. Problems of these form have been studied in the fields of *decentralised control* (Šiljak 1991, Sandell, Varaiya, Athans & Safonov 1978) and *game theory* (Basar & Olsder 1995) and applied to large-scale industrial problems (Samyudia, Lee & Cameron 1994, Samyudia, Lee, Cameron & Green 1995, Guo, Hill & Wang 2000). The lack of communication naturally results in a suboptimal global performance.

This can be improved if the agents can be allowed to communicate. In the setting of dynamical control, the agents can be made to communicate various forms of information, for example: (a) the abstraction of their local dynamic models, (ii) the predictions of their future interactions, (iii) the *cost-to-produce* and *cost-to-respond* to incoming and outgoing interactions, *etc.* (Tenney & Sandell 1981). With increased availability and reliability of communication tools, such communication based structures, in particular those based on prediction, have found application in distributing so-called model predictive control calculations (see, for example, Camponogara *et al.* 2002, Venkat *et al.* 2006, Keviczky, Borrelli & Balas 2006).

A large body of work on distributed algorithms that also uses communication as part of problem solving belongs to so-called *relaxation* techniques from optimisation and operations research literature (Bertsekas & Tsitsiklis 1989). In simple terms, the relaxation methods build upon a principle that, if problem structure permits, the optimisation step in a centralised technique, *e.g.*, a gradient step $x(t + 1) = x(t) - \gamma \nabla F(x(t))$, can be split and distributed among agents responsible for subsets of variables. The agents iteratively solve their local problems and communicate these local solutions in some form. The overall solution is made to converge by imposing a global constraint such as the order in which their local problems are solved. See (Bertsekas & Tsitsiklis 1989) for an extensive overview of this class of algorithms. The concept of dynamic programming also provides a communication-based method for solving multi-stage problems such as in process synthesis (Jackson 1964b, Jackson 1964a, Rudd & Watson 1968) and process modelling (Kisala, Trevino-Lozano, Boston, Britt & Evans 1987, Westerberg, Hutchison, Motard & Winter 1979, Alkaya, Vasantharajan & Biegler 2000).

An important class of distributed solution techniques based on so-called *nested decomposition* concept have remained dormant over the years (Glassey 1973, Ho & Manne 1974, O'Neill 1976), however, as shown later in this monograph, these can provide an excellent tool for solving distributed control problems arising in multi-stage networks such as process plants. The word *nested* refers to a sequential solution of multiple, two-level coordination problems, each associated with a junction (or link) connecting two or more agents or subsystems. Starting from the root of the network, each agent in the sequence coordinates its own actions plus those of its predecessors and passes relevant information down to its successors. The interactions repeat across the network whereby agents incrementally build and refine abstractions of cost objectives and feasible regions and utilise this information in solving the global problem. See Chapter 6 for further details on nested decomposition.

Discussion

Both coordination methods described above offer improved benefit of reconfigurability over conventional methods because the formulation of local controller problems are distributed and can be easily modified. However, both coordination methods also need a separate mechanism for coordinating the local solutions to guarantee coherent global operations. Historically, coordination is perceived as a complex process difficult to implement within industrial process control due to: (a) the process problems can be complicated due to the use of material and energy recycles and (b) the problem formulations used at higher-levels, *e.g.*, in planning and scheduling problems, remain generally monolithic. The use of coordination in this context for problem solving can lead to slower convergence and may not work reliably due to the reliance placed on communication tools. However, with the advances in communication and computing technologies in recent years, these issues have remained less of a concern nowadays. As discussed earlier in the previous section, if the complexities of recycles and heat integrated are treated secondary to the reconfigurability of operations then the benefits offered by coordination methods, in particular those based on distributed coordination, can provide attractive alternatives for building modular control architectures that also support such rapid integration and reconfiguration (Backx *et al.* 2000, Samad *et al.* 2007).

2.4.3 Distributed Paradigms for Reconfigurable Manufacturing Control

As mentioned earlier, distributed approaches have been used previously in developing greater reconfigurability in distributed manufacturing control. The driver for such development was the business pressures felt by manufacturing industries in the early nineties. The increased attention on product customisation and diversification led to many researchers tackle the problem of manufacturing agility by seeking inspiration from other man-made or natural systems where adaptability to change has been key to their survival. Some examples of new paradigms include *fractal factory* (Askin, Ciarallo & Lundgren 1999), *bionic manufacturing*, (Ueda 1992, Tharumarajah, Wells & Nemes 1996), *holonic manufacturing* (Christensen 1994, Seidel 1994) *etc.* Although motivational and insightful, many of these new approaches failed to make an impact due to their rather radical nature. The two concepts which did succeed namely, holonic and agent-based manufacturing, led to major research interests internationally. We give in this section a brief overview of the research in these fields with an aim to identify the background concepts relevant to this work. More comprehensive overviews can be found in surveys (McFarlane & Bussmann 2000, Mařik, Fletcher & Pěchouček 2002, Babiceanu & Chen 2006, Shen, Hao, Yoon & Norrie 2006, Shen, Wang & Hao 2006). Industrial deployment of these technologies has been reviewed in (Mařik & McFarlane 2005, Pěchouček & Mařik 2006).

Holonic Manufacturing Systems

The concept of *holon* was proposed by Koestler (1967) in his studies on the evolution in biological and social systems. The word holon, a combination of *holos* (meaning 'whole') and *-on* (meaning 'part'), describes a self-reliant element of a system that is able to exist on its own as an autonomous entity and also is able to integrate with other such elements in the system to create a larger system *i.e.*, a holon demonstrates the dual characteristics of *autonomy* and *co-operation* at the same time. The holonic concept was brought to manufacturing by Suda in his work (Suda 1989, Suda 1990) where he observed that properties analogous of holons in a biological or social system would be desirable in a manufacturing environment when faced with the challenges of customisation and global competition. To motivate the analogy, he proposed the concept of *manufacturing holons* and the associated manufacturing model as *holonic manufacturing systems*. Suda's work led to a number of research efforts promoting the holonic concept as the paradigm for next generation manufacturing systems. The motivation behind these developments was to create a distributed manufacturing architecture that is made up of a modular mix of (semi-)autonomous manufacturing holons that can make stand-alone decisions and are able to collaborate among themselves to produce goods. A bottom-up integration of manufacturing holons, achieved through reconfigurable, distributed interactions is then considered a rational approach to building manufacturing systems of the future.

Agent-Based Manufacturing Control

In parallel to holonic research, the concept of agent-based control also emerged as a paradigm to address similar challenges in manufacturing. An agent, by definition, is a flexible, computational element possessing a level of intelligence to operate independently (Wooldridge 2002). A multi-agent system, comprising multiple interacting agents, is considered to provide the *intelligence* necessary to create a dynamically reconfigurable and to an extent self-organisable design of manufacturing elements.

The agency concepts, while studied previously in computer science, were largely untested in manufacturing and led to bringing together the researchers from holonic and agent communities, with the former providing a physical platform for building agent-based manufacturing systems (Fischer 1999, Mařík *et al.* 2002, Giret & Botti 2004). The concepts of *pro-activeness* and *reactiveness* from agency research are since used widely in holonic and agent research to define the various coordination issues such as communication protocols, decision-making strategies and the planning and scheduling algorithms (Mařík *et al.* 2002, Shen, Hao, Yoon & Norrie 2006).

Holonic and Agent Research in Discrete Manufacturing

The mainstream holonic or agent research, while focussed on discrete manufacturing, has followed the so-called *low and late commitment* principle from the theory of flexibility (Valckenaers & van Brussel 2005), which suggests that to enhance flexibility a designer should commit to a design decision as late as possible and the severity of the commitment should be kept as low as possible, *i.e.*, (a) where possible, the design decisions should be postponed or avoided by providing alternatives and (b) the design process should avoid building "inertia" that makes it harder to rectify the errors at a later stage (Wyns 1999).

In a make-to-order environment, the principle of late commitment has been employed to provide the support for customisation and diversification of products. The concepts of so-called *product holon* and *resource holon* are introduced – the former representing the recipe knowledge on 'how to produce a product' for a specific order and the latter as the production capabilities in terms of machines and other resources available on the shopfloor (van Brussel *et al.* 1998, Chirn & McFarlane 2001, Leitão & Restivo 2006). These two aspects are separated in the design and only integrated during run-time operations via distributed interactions between product and resource holons. By delaying their integration, the developers of the recipe knowledge or the machine control receive a freedom to choose design solutions that best suit the local conditions. Equally, the most recent status of conditions on the shopfloor is taken into account before assigning tasks that fit with the order requirements. As a result new orders can be dynamically introduced or the existing orders shuffled to better utilise the resources.

The principle of low commitment is also extended to engineering and design of control system so as to suggest a method of *top-down decomposition, bottom-up integration*. A bottom-up method is preferred for integration as it avoids the pitfalls of initial global design which can be restrictive (van Brussel *et al.* 1999). In the proposed method, the decomposition of end-user requirements still occurs top-down however little or no design choices are made enroute. Resulting bottom-level requirements from the decomposition are then associated with appropriate holons from a set of pre-identified holon types. Selected holons are then designed and implemented in a bottom-up manner such that their final designs are reusable, preferably of multifunctional nature. To support the identification of holons, a number of different classifications have been suggested in the form of so-called *reference architectures*. Some prominent examples of these include PROSA (van Brussel *et al.* 1998), HCBA (Chirn & McFarlane 2001), ADACOR (Leitão & Restivo 2006) and Meta-Morph (Maturana & Norrie 1996, Shen, Maturanan & Norrie 2000). Internal design of holons that supports this architectural research has also received vivid interest. Some key references include (Christensen 1994, Rannanjärvi & Heikkilä 1998, Heikkilä, Järviluoma & Juntunen 1997, Fischer 1999, Brennan, Fletcher & Norrie 2002).

The holons operate in a distributed mode and share information to reorganise their operations and coordinate associated decisions. The functionality of conventional hierarchy is loosened and distributed among holons; holons solve related planning, scheduling and control problems in a distributed form. Development of coordination techniques to support these interactions has formed an essential part of research, not just to define the problem solving mechanisms but also to provide an ontological description of the interactions that are used to standardise the communication protocols used by holons and their internal designs. The key solution concepts considered include contracting (Smith 1980), lagrangian decomposition (Gou, Luh & Kyoya 1998), market programming (Váncza & Márkus 2000) and behaviour-based techniques (Valckenaers, van Brussel, Kollingbaum & Bochmann 2001, Tharumarajah & Wells 1996). Associated applications in control cover holonic planning (Deen 1993), scheduling (Gou et al. 1998, Sousa & Ramos 1998) and execution control (Heikkilä et al. 1997). See (McFarlane & Bussmann 2000, Tharumarajah 2001, Shen, Hao, Yoon & Norrie 2006) for recent overviews.

Holonic and Agent Research in Process Applications

Research on holonic or agent-based based systems or similar principles has been scarce in the process industry. One of the early interests was in agent applications to support design and engineering of process plants purely to perform mundane tasks such as collecting the data and checking different design alternatives. (Jennings, Faratin, Norman, O'Brien, Odgers & Alty 2000, Batres, Asprey, Fuchino & Naka 1999). More technical use of agents has been found in distributed fault diagnosis (Seilonen, Appelqvist, Halme & Koskinen 2002, Eo, Chang, Shin & Yoon 2000, Maturana, Tichý, Slechta, Staron, Discenzo & Hall 2003). The agents here represent and monitor one or more pieces of equipment. During a fault scenario, they build and postulate possible hypothesis of the fault scenarios and communicate results to eliminate unlikely possibilities. Ultimately they recognise the nature and extent of the fault and advise the operator of potential remedies for repair. On a different front, Chokshi, Matson & McFarlane (2000) considered a holonic framework for batch re-scheduling in a steel-making. The concept of *partial global planning* (Durfee & Lesser 1991) was considered from distributed AI research to define the coordination of start and end-times of batch tasks and the movement of *ladles* between unit operations.

More recently, agent-based research has found a surge of interest in the coordination of process supply chains. Among them the key references include (García-Flores, Wang & Goltz 2000, Julka, Srinivasan & Karimi 2002, Julka, Karimi & Srinivasan 2002, Gjerdrum, Shah & Papageorgiou 2001, Aldea, Bañares Alcaántara, Jiménez, Moreno, Martínez & Riaño 2004). Backx et al. (2000) gave an interesting insight on the need for *intentionally dynamic, supply-chain conscious* process operations. They showed that a decentralised design of process plants operating in a so-called *cooperative* mode will be essential to support the future requirements. Their initial results defining the control algorithms for market-oriented optimisation and scheduling of process operations are reported in (Tousain 2002, Tousain & Bosgra 2006).

Diversity	Modifiability	Responsiveness	Fault-tolerance		
✓	✓	✓	✓	• Modular, multipurpose, re-usable design	**Architectural Properties**
✓	✓	✓		• Separation of product recipe information	
✓	✓		✓	• Top down decomposition, bottom up integration	
✓	✓	✓	✓	• Low and late commitment	**Operational Properties**
✓		✓	✓	• Dynamic integration of product recipe information	
✓		✓	✓	• Distributed decision-making and control	
✓		✓	✓	• Proactive and reactive behaviour	

Fig. 2.8. Satisfaction of reconfigurability requirements using a distributed approach (the shaded labels show a major link)

2.4.4 Summary

Fig. 2.8 summarises the key properties of distributed approaches in holonic and agent research by linking them with the reconfigurability requirements in Fig. 2.7. As can be seen, the architectural properties can address the static requirements of product/process diversity and easy modifiability, while the operational properties can address the dynamic requirements of responsiveness and fault-tolerance and also help improve the diversity via dynamic integration of product information.

2.5 Reconfigurable Control Research in Other Domains

The concept of reconfigurable control based on distributed approaches has also been studied in domains other than manufacturing, particularly where it remains impossible to employ a centralised control structure. A brief review of this related research is presented in this section to gain insights on the nature of approaches used therein to attain reconfigurability.

2.5.1 Formation Control of Robots or Aircraft

Maintaining a formation of multiple robots or aircraft operating in a close proximity has gained interest recently in areas where unmanned operations are essential (Egerstedt & Hu 2001, Beard, Lawton & Hadaegh 2001, Giulietti, Pollini & Innocenti 2000). Typical of such applications include exploration of unknown environments, coordinated path following and pushing objects in a coordinated fashion. The formation may be time-varying and may be subjected to various hard or soft constraints, such as retain minimum distance between robots or aircraft.

The use of multi-agent control schemes based on coordination have become popular in this domain primarily because the environmental stimulations in which the distributed entities operate remain unknown *a priori*. Beard *et al.* (2001), for example, classified the coordination approaches used into three categories: (i) *leader-following*, where all agents (*i.e.*, robots or aircraft controllers) follow the path of a common leader agent; (ii) *behavioural*, where the group behaviour emerge from the localised behaviour of all agents and (iii) *virtual structure*, where the formation is treated and controlled as a single structure, which in turn directs the actions of the individual agents. See (Beard *et al.* 2001) for further details.

2.5.2 Congestion Control in Communication Networks

With ever increasing use of internet and communication technology, the control of traffic management in communication networks has become important. The problem is further complicated because of uncertainties in the time at which traffic may arise or the amount of network resources that it may demand (Kelly, Maulloo & Tan 1998). One problem in traffic management is *flow control* – for a given network configuration, adjust the incoming traffic such that the network utilisation is maximised. The other problem is *routing* – for a given network configuration and utilisation level, determine the routing of data packets across the network such that the priority constraints (*e.g.*, importance of certain data over others) are satisfied.

Two streams of solution strategies have evolved over the years for these two problems. One stream assumes that individual users are self-maximising agents and aim to maximise their utility for a given shared access of the network. The concept of non-cooperative game theory (Basar & Olsder 1995) is used to characterise the resulting equilibrium conditions for the solution. The properties such as fairness, efficiency of utilisation and quality of service are studied here (Korilis & Lazar 1995, Korilis, Lazar & Orda 1997, Altman, Başar & Srikant 2002, Orda, Rom & Shimkin 1993). The other stream takes a control-theoretic view where the aim of the study is the stability of the equilibrium in the presence of feedback delays arising between user/source pairs. The metrics such as convergence, capacity tracking and robustness to changing dynamics are studied to define the distributed control laws for traffic management (Kelly *et al.* 1998, Vinnicombe 2000, Johari & Tan 2001).

2.5.3 Power Systems and Electricity Markets

Increasing competition has led to many electricity markets being deregulated worldwide. Under new trading rules, individual generators and consumers submit their bids for supply or demand of electricity to a common regulator. The regulator evaluates the bids based on forecast demand and decides a market clearing price at which the electricity is traded. Since the generators and consumers operate as self-utility maximisers, the concept of non-cooperative game theory provides a right platform to study the equilibrium pricing and establish trading mechanisms to reach equilibrium for a given structure of grid and transmission capacity (Stothert & MacLeod 2000, Green & Newbery 1992, Kleindorfer, Wu & Fernando 2001, Hobbs, Metzler & Pang 2000).

Power networks also face the problem of responsiveness against faults and disturbances. Similar to process plants, the grid connections between individual generators or consumers introduce tight coupling between their local processes. A minor or small fault in one part of the network can, as a result, propagate to other parts or the whole network if not properly managed in time. A prompt diagnosis and isolation of fault thus remains ever so important, but the ability of remaining generators or consumers to compensate for this grid imbalance also is equally important to avoid blackouts. It is however impossible to manage this problem centrally due to large size of the networks in most cases. Instead, the concept of decentralised control has been used frequently as discussed earlier in the review of distributed coordination literature (see, for example, Guo *et al.* 2000).

2.5.4 Supply Chain Management

Research in supply chain management and control has flourished in recent years due to increased attention on customisation and diversification in global markets (Maloni & Benton 1997, Tayur *et al.* 1999, Strader, Lin & Shaw 1998). The supply chains nowadays are required to respond and adapt to constantly changing conditions. Their conventional monopolistic form cannot however realise this level of response due to fixed and rigid structure. Instead supply chains are now regarded as supply chain networks (SCN) – an integration of multiple supply chains that evolve and scale according to changing needs of the market (Fox, Barbuceanu & Teigen 2000).

Specifically, the concept of so-called *virtual enterprise* (Strader *et al.* 1998, Camarinha-Matos *et al.* 2003) has emerged. In a virtual enterprise multiple equal-interest companies come together to form a chain that can exploit the fast-changing market opportunities. Each alliance is formed and operated via distributed interactions between companies. Once the opportunity ceases, the alliance is dissolved and the companies move towards forming new partnerships. The effective operation of supply chains, in particular virtual enterprises, requires sharing information between partners and synchronising their local operating policies. The multi-agent technology in this sense has provided a platform for modelling the underlying distributed interactions. See (Chaib-draa & Müller 2006) for a collection of recent references.

2.5.5 Discussion

When considering reconfigurable process control, many lessons can be learnt from developments in the above and other domains. Similar to process plants, in all four domains described above the agents are subjected to hard or soft network constraints. In formation control, robots or aircrafts must maintain a fixed distance. In communication networks, the capacity of network links may limit the data that the users can put on the network. In supply chains, companies remain connected via transport routes and the operating policies they use also need to fit with those of immediate customers, suppliers and transporters. Similarly, in all four domains, the agents must also maintain a stable operation of the global system under time-varying conditions. In formation control or supply chains, the behaviour emerges via co-operation between distributed entities, while in communication or power networks this is enforced by the need for reaching a system-wide equilibrium. Note that in all four domains these static or dynamic properties of the global system emerges via direct, bottom-up interactions between distributed agents.

The research in supply chain networks is particularly relevant to this work. Supply chains exhibit a multi-stage character of commodity flow which – in a sense – is similar to the flow of materials in manufacturing systems comprising network constraints such as process plants. The notions such as 'product', 'product demand', 'customer order' as viewed in a manufacturing system also relate to supply chains in a similar manner. Interestingly, the supply chain paradigm also extends the market or contracting approach used in previous holonic or agent research by introducing the network interactions of 'supplier-to-supplier' type apart from 'customer-to-supplier' type in a market or contracting approach. As discussed in the next chapter, this extension provides the basis for our distributed approach to reconfigurable process control.

2.6 Summary

This chapter has built a foundation for understanding existing work in process control, distributed control and coordination and the role of reconfigurability in this domain. We next move onto the main body of the monograph which proposes a distributed approach to reconfigurable process control.

Part II

A Distributed Reconfigurable Process Control Approach

3

DRPC: Distributed Reconfigurable Process Control

3.1 Introduction

In the previous chapter we noted that a distributed approach to the design of a process control environment may provide a route to increased reconfigurability and with that the associated business benefits. In this overview chapter and the following three chapters we begin to build up a blueprint for how such a distributed control system might be constructed. We have already noted that such an approach is – conceptually – fundamentally different to conventional, hierarchically-based control systems and because of this we begin with a re-casting of the process control system structure before moving on to examine the way such a system might operate.

In this chapter we first gather together the needs for a reconfigurable process control system – as discussed in Chapter 2 – and then marry these with the notions of distributed coordination as defined from the related but different developments described in Section 2.5. This allows us to produce a conceptual overview of how a distributed and reconfigurable process control system might operate. From here, we identify the key developments needed in order to construct such a system, namely, the architecture and element designs, the interaction between distributed elements and the governing optimisation strategy for achieving globally well behaved control.

3.2 Addressing the Needs for Reconfigurable Process Control

In the solution we are going to develop we seek to address the needs for reconfigurable process control as identified in Section 2.3.3. These were summarised into four system requirements: (i) product and process diversity, (ii) easy modifiability, (iii) responsiveness to change and disturbances, and (iv) fault-tolerance with graceful degradation of performance (Fig. 2.7). The state-of-the-art review of research in process control showed that each of these requirements have been addressed – but only individually and only for rather limited class of applications. A holistic approach addressing them within a single framework is yet to appear. We propose here that a distributed coordination approach based on holonic principles can provide one such approach. The previous results in holonic research, however, apply to discrete manufacturing and therefore do not translate straight to process control. So, to develop a distributed RPC approach, we need to understand what opportunities and constraints exist in process control that are different to discrete control, and from that, decide how to adapt the existing holonic research.

3.2.1 The Reconfiguration Process

Before assessing how to use the existing research to address the needs of reconfigurability, we firstly examine the process of reconfiguration itself. Fig. 3.1 outlines all the key steps in the reconfiguration of a complex system which range from the identification of an opportunity to the coupling/recoupling of elements, the reorganisation of those elements and the monitoring of the outcome. First we concentrate on the central section: that of effecting the reconfiguration process. Intuitively, this can be split into three actions: (i) *decouple* – pull apart system elements from existing configuration, (ii) *reorganise* – reorganise them into new configuration, and (iii) *recouple* – put them together to operate. All three actions may involve a physical change (physical set-up of the process units or their interconnections) and/or a control change (operating settings, recipe parameters *etc.*). The reconfiguration may be requested as planned (*e.g.*, introduction of a new product order) or unplanned (*e.g.*, failure of a process unit). An RPC system capable of tackling these conditions shall provide the support necessary to carry out the above actions smoothly and efficiently, and preferably in an *automated* fashion. In addition, the RPC system must also *identify* that a reconfiguration is necessary from *monitoring* of the plant conditions or the arrival of new production opportunities (new orders *etc.*), and *define* the structure of the new configuration. In conventional practices, these actions would be performed offline (when planned) or by human intervention (when unplanned). However, in more dynamic scenarios that can arise in future, such methods may fail to cope with the complexity and fast timescales that may be demanded. Instead, we believe these actions shall as well be included as part of the scope of the RPC system with, if necessary, an extra level of automation and intelligence provided that induces self-reconfiguration.

3.2.2 Adapting Existing Holonic Systems Research

The previous holonic research has used distributed interactions between holons to manage the reconfiguration process in a manner described above.

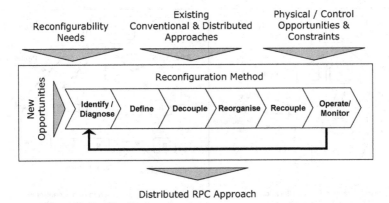

Fig. 3.1. Research approach

Generally, a contracting protocol (Smith 1980) or its extended variants such as based on lagrangian decomposition (Gou *et al.* 1998) or market programming (Váncza & Márkus 2000) are used to define the information structure for interactions between holons (Fig. 3.2(a)). In contracting the interactions occur between product and resource holons. In a so-called *task-based view* of contracting, the product holons act as the *managers* and distribute tasks to appropriate resource holons. In turn, they also coordinate the flow of parts across shopfloor. In a dispatching mode of operation, the interactions build progressively with new tasks only assigned when the previous tasks have finished. The resource holons themselves are assumed physically decoupled (in jobshops) or connected via buffers such as storage units and conveyor queues that are assumed to decouple their operations (in flowshops). The resource holons therefore do not interact directly or coordinate their operations. In an alternative *resource-based* view, the resource holons instead act as the managers and announce their availability and accept tasks that best suite the local conditions. Again the interactions occur between product and resource holons with the former acting as the coordinators of the flow of parts.

In a continuous process the situation remains different though because the process units remain tightly connected via piping streams and their local dynamics interact in most cases due to lack of interim storage or buffer tanks. That means, the processing tasks assigned to any process unit must match – in a physicochemical term – to that of its neighboring units. A tight coordination of unit operations is thus essential, both at the time of allocating tasks to process units and also when these tasks are being executed. In most cases it may also be necessary that the entire task sequence in the product recipe, *i.e.*, from raw-materials to end-products, is developed first before the execution of the first task can start; a dispatching mode of task release and assignment cannot simply work. In these conditions if the above task-based view is used for managing interactions, then it is likely that the coordination

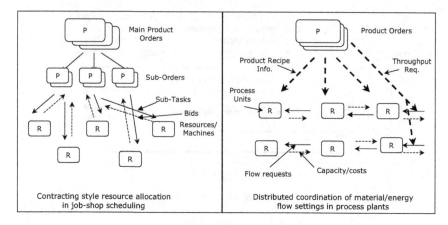

Fig. 3.2. Distinction between: (a) contracting-style resource allocation in discrete manufacturing, (b) proposed form of distributed coordination in process plants

effort required to satisfy these additional constraints can become excessive due to centralised role of product holons and the chances of potential conflicts between their task assignments. Instead, a distributed method for RPC must be better off by achieving coordination through direct interactions between production functions themselves (*i.e.*, resource holons) whilst also interacting with product functions (*i.e.*, product holons) where necessary. This distinction is clarified in Fig. 3.2(b).

3.2.3 Analogy from Virtual Enterprise Management

A useful coordination method for a DRPC approach can be developed by extending the contracting principle in previous research with an analogy from supply chain management. In particular, we borrow an analogy from so-called *virtual enterprises* or *dynamic* supply chains as discussed earlier in Section 2.5. In a virtual enterprise multiple equal-interest companies come together to form a temporary alliance that delivers the fast-changing customer demands. The alliance evolves in time and adapts to changing marketplace when necessary. The reconfiguration of the whole chain, including that of coordinating the material flows, occurs via direct, distributed interactions between companies themselves. A framework analogous to these interactions of companies can be considered to define the interactions of product and production functions within an RPC system.

In particular, we consider an analogy from the 'life cycle model' of a virtual enterprise. A virtual enterprise normally goes through four main phases during its life cycle as shown in Fig. 3.3: (i) identification, (ii) formation, (iii) operation, and (iv) termination (Strader *et al.* 1998). The identification phase starts with a research of available market opportunities and identifying an opportunity that can be pursued further. This is then input to the formation

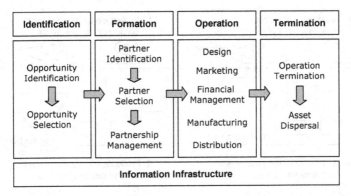

Fig. 3.3. Life-cycle model of a virtual enterprise (*Source:* Strader *et al.* 1998)

phase. The formation involves identification and selection of partnering companies and building from that a chain that can deliver the order requirements. Different combinations of partners and business processes may be evaluated before the companies arrive at a choice of the configuration. The selected partners then integrate their business processes during the operation phase in order to deliver the order. During this they may exchange local information such as stock levels, demand forecasts *etc.* to improve their visibility across the chain. The network is finally dissolved and the shared assets, if any, are dispersed or re-used to initiate a new opportunity.

In an analogous manner, the production functions in the proposed DRPC approach use distributed interactions to manage the reconfiguration process in Fig. 3.1. An outline of how this analogy would operate is illustrated in Fig. 3.4, which compares the order fulfillment process in a virtual enterprise and in a DRPC system. In a virtual enterprise, the evolving market opportunities drive the partnering companies to come together and form alliances while in a DRPC system, the customer orders (or so-called *product elements*) drive the different production functions to configure appropriate process schemes and deliver the order requirements.

The analogy, while conceptually sound, is faced with certain challenges. Process plants, unlike supply chains, are characterised by shorter time-scales, the non-linear and dynamic behaviour of process units, and the material and energy recycles – the features which are not normally critical in supply chains. However, if these complexities are taken aside and if the processes are seen purely as chains of material and energy flows, then it is possible that the network behaviour of process units, *e.g.*, in a multiproduct or multipurpose plant, can still be examined in a manner similar to that of companies in a virtual enterprises. The analogy in this sense can provide an useful aid to visualise the operations of distributed production functions in a process plant in a manner much similar to the use of contracting in previous holonic research.

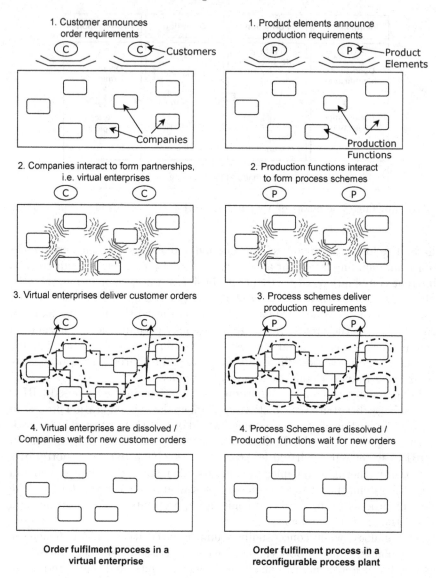

Fig. 3.4. Order fulfillment process in a virtual enterprise and a DRPC system

3.3 Introducing the DRPC Approach

In line with the previous research in holonic and agent-based industrial control, we now begin to develop the DRPC approach by describing the tools necessary for constructing an RPC system. The method of *top-down decomposition* and *bottom-up integration* is considered the key to these developments. In particular in the following chapters we focus on the following three areas.

i. *Distributed Control Architecture:* A control architecture is developed in order to characterise the different production functions (so-called *process elements*) and their primary control responsibility.

ii. *Distributed Interaction Model:* Next the structure of information exchange between process elements is defined so as to cover the interactions necessary for implementing the reconfiguration process in Fig. 3.1.

iii. *Distributed Control Strategy:* To support their reconfiguration decisions, the process elements are finally supplied with control strategies in the form of a distributed algorithm. Only a generic problem is investigated at this stage so as to complete the architectural description.

The following chapters in this part of the book address each of these developments in turn.

4

Reconfigurable Process Control Architecture

4.1 Introduction

We now begin to construct the DRPC system from its basic elements and specify how these elements are defined for typical process control functions. The connection between these elements – the so-called control architecture – defines the structure for the process control system resulting from their combination.

4.1.1 Overview

We start with understanding the terms *architecture* and *control architecture*, *i.e.*, what properties should a control architecture have and what are its key elements.

The term *architecture* can be defined broadly as the attributes of a system as seen by its designer, or formally as, a conceptual structure of the system that also defines its functional behaviour while being distinct from the detailed design and physical implementation (Amdahl, Blaauw & Brooks 1964). An architecture in this sense forms a critical input to the design process to lay down the specifications of end-user requirements based on which the actual system can be built.

Over the years, two different meanings of 'architecture' have evolved in systems engineering: (i) the architecture as a generic 'style' or a 'method' for building one or more systems, called the *reference architecture* and (ii) the architecture as a 'product' or a 'template' for a specific system, called the *system architecture* (Williams 1989, Zwegers 1998). The reference architecture sets out the generic behaviour and attributes and possibly the rules of design for a number of similar systems. A system architecture instantiates the behaviour and attributes of the reference architecture by applying these rules to a specific application. Fig. 4.1 outlines the use of reference and system architectures within overall systems engineering process.

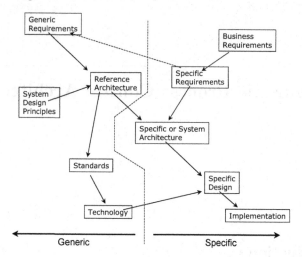

Fig. 4.1. Role of architectures in systems engineering (*Source:* Williams T.J. 1989)

The term *control architecture* refers to an architecture of a manufacturing control system (or in the context of this research, for a process control system). We limit it to be a reference architecture in this text. The role of a control architecture in this regard is to allocate the various decision-making responsibilities for production control to the specific control components or controllers. Further, it should determine the relationships between these controllers so as to establish a mechanism for coordinating the execution of their decisions (Dilts, Boyd & Whorms 1991, Senehi, Wallace & Luce 1992).

The research on manufacturing control architectures has evolved over the years. Historically, the early architectures defined as part of Computer Integrated Manufacturing (CIM) were hierarchical – so-called 'proper' hierarchical as Dilts *et al.* (1991) call them. Some key examples include AMRF (Jones & McLean 1986) and MSI (Senehi *et al.* 1992) in discrete manufacturing and Purdue Reference Model (Williams 1989) in process industry. Hierarchy helped manage the size and complexity of control functions that the earlier centralised structures failed to handle. But as the time progressed, it was recognised that hierarchy can have its own shortfalls. The inflexible structure of hierarchy due to multiple levels of control and the delays in passing information between these levels could result in poor response to unforeseen change and disturbances. To overcome these shortfalls so-called *heterarchical* or flat architectures were proposed as comprising distributed, locally autonomous controllers without any master/slave relationships (Duffie & Piper 1987, Duffie & Prabhu 1994). The benefits of hierarchy and heterarchy have been long debated as heterarchy can result in chaotic performance due to lack of coordination where hierarchy was shown to perform better (Bongaerts, Monostori, McFarlane & Kádár 2000). *Holarchy*, a term coined as part of holonic research

(Koestler 1967, Christensen 1994), is considered to deliver the benefits of both hierarchy and heterarchy whilst also avoiding their shortcomings. Unlike hierarchy, the system elements in a holarchy remain distributed, loosely-coupled, but unlike heterarchy they also coordinate their operations across the plant. They can behave both as pro-actively and reactively under different conditions that in a way enhances their reconfigurability. We exploit this dual property of holarchy in defining the behaviour of process elements in the control architecture to be developed in this and later chapters.

4.1.2 Requirements for the RPC Architecture

In this chapter, we aim to develop a control architecture for RPC systems with a focus on so-called semicontinuous class of process systems. The architecture is expected to help at least meet the requirements from Fig. 2.7 of product/process diversity and easy modifiability as they both heavily depend on the architectural properties of a process control system. In addition, it should help improve responsiveness and fault-tolerance of the system by ensuring that constituent control elements of the architecture are sufficiently decoupled and that the propagation of disturbance or failures across the system remains limited or occurs gracefully.

This chapter is structured as follows. Section 4.2 next describes the structure, data models and basic control functions of distributed process elements in the architecture. An incremental approach to migrating to this fully distributed form of control is suggested in Section 4.3 so as to allow industrial practitioners to experiment with these new concepts using existing off-the-shelf control tools. Section 4.4 applies the architecture to an example polymer process plant. Some comments on the architecture in terms of the above mentioned requirements and the other conventional and distributed architectures are presented finally in Section 4.5.

4.2 Specification of Process Elements in a RPC System

We now describe the specification of distributed process elements in the proposed RPC architecture. Following the approach described in Section 3.2, we consider a supply chain (in particular virtual enterprise) based analogy to visualise the structure and behaviour of elements in the architecture.

4.2.1 Basic Types of Process Elements

The proposed architecture divides the functionality of a process control system into four primary types of process elements, called: (i) unit element, (ii) piping header element (in short, header element), (iii) service supplier element (in short, service element),[1] (iv) product element. The functionality is divided based on physical structure of the process instead of temporal or multi-level decomposition as in a hierarchical system.

The functionality of these individual types of process elements in the architecture can be defined as follows:

- *Unit Element:* A process unit element (in short unit element) represents a physicochemical processing task such as reaction, distillation *etc.* in the process. The task may have associated with at least one but possibly more control decisions that the unit element can regulate on its own.

- *Header Element:* A header element represents the logistics of materials or services within a specific segment of the overall process network. Physically, it may contain a number of piping streams, transfer equipment (pumps, compressors *etc.*), final control elements, energy transfer units (heat exchangers *etc.*) and storage units. These component sub-units should not incur any physicochemical operation, however they can be used to change the physical state of the material or service being transferred, *e.g.*, heat, cool, pressurise or depressurise them.

- *Service Element:* A service supplier element represents a custodian responsible for allocating a service to customer process elements that use this service in their local tasks. The customer elements can be either unit or header elements. Multiple service elements may exist in the process, each supplying one or more different services.

- *Product Element:* A product element represents the production requirements of a specific customer order in the form of a product recipe (specifying the sequence of processing tasks to be used or allowed) and other requirements such as quality, quantity and throughput of the product demand. Multiple product elements may co-exist in the process, each representing a specific customer order, but only a few may be produced at a time. Note that unlike the previous three elements, the product element does not have a physical presence in the process; it only acts as an information component supplying necessary product information to other process elements.

Fig. 4.2 depicts examples of various unit, header and service elements that can be found in process industries. Important to notice is that the header elements decouple the operations of unit and service elements in a sense that the physicochemical tasks of unit elements or the service supply tasks of service elements can be identified and defined more clearly and separately from the

---

[1] The term *service* refers to utilities such as steam, cooling water, electricity and other such enabling facilities (for example manpower) that are used in the execution of various processing tasks and the transfer of materials.

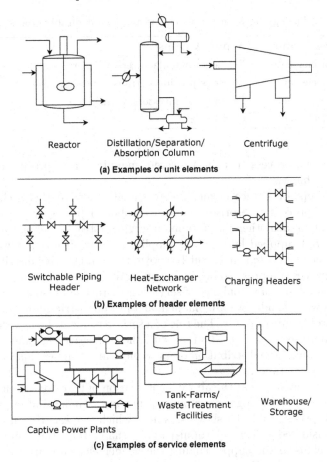

Reactor   Distillation/Separation/   Centrifuge
           Absorption Column

(a) Examples of unit elements

Switchable Piping   Heat-Exchanger   Charging Headers
Header   Network

(b) Examples of header elements

                     Tank-Farms/
                     Waste Treatment   Warehouse/
                     Facilities   Storage

Captive Power Plants

(c) Examples of service elements

Fig. 4.2. Examples of unit, header and service elements in process industry

transfer/transform tasks of header elements; the header elements can thus be made flexible as and when necessary by adding extra transfer facilities without having to modify the interface of unit or service elements.

We note that the above identification of such element types is not strictly new to the distributed coordination field. Except for the service element, the notions of unit, header and product elements have previously appeared in vivid forms as so-called resource, transport and product holons in other distributed architectures in holonic and agent research (*e.g.*, PROSA (van Brussel *et al.* 1998), HCBA (Chirn & McFarlane 2001), HSCF (Cheung, Yeung, Ng & Fung 2000), ADACOR (Leitão & Restivo 2006)). However, as explained later in this chapter and the next chapter, the roles and interactions of process elements are different in the current architecture than these previous architectures. The differences primarily emerge due to the physically distinct nature of operations

Table 4.1. Analogy between supply chains and reconfigurable process plants

Process Plants	Supply Chains
Unit elements	Echelons (manufacturers, retailers *etc.*)
Header elements	Logistics providers (transporters, storage units *etc.*)
Service elements	Facilitators (investors, banks *etc.*)
Product elements	Customers

in a continuous process then in a discrete manufacturing process discussed later in this chapter.

The concept of service element is specifically new to this architecture. It relates to process enterprises where a number of plants or process units situated next to each other share common services (steam, cooling water, raw-materials, *etc.*) supplied by separate supplier facilities (captive power plant, cooling water plant, *etc.*). It is widely known that an effective distribution of common services can prove to be significant at times when the supplier plants fail or the supply-demand balance of services is disturbed for some reasons. At other times when conditions are planned, an optimal distribution can increase company profits substantially. The role of a service element, being a custodian of one or more services, is to interact with the respective customer elements so as to coordinate the distribution of its services in a manner that is effective and responsive at times.

We note in passing that the above identification of four process elements is also related to their analogous components in supply chains and in particular virtual enterprises. Table 4.1 shows this link, which suggests that if a process plant is considered a form of (mini-)supply chain, then the unit elements are the echelons in the supply chains, the header elements are the logistics providers, the service elements are the facilitators or service providers, and the product elements are the final customers. The analogy thus provides a systematic, ontological concept (as an extension to the contracting principle in previous holonic or agent research) to define the interactions of process elements. This is discussed in more detail in the next chapter.

4.2.2 Data Model and Control Functions of Process Elements

All four process element types possess associated roles, data models and control functions in the architecture. These can be described as below. This information then forms a part of the interactions of elements in the next chapter:

- *Unit Element:* The role of a unit element is to perform one or more processing tasks. To satisfy this role, it executes the following functions: (i) identify the processing tasks it should perform by interacting with respective product elements and other unit elements; (ii) acquire necessary feedstocks and services for these tasks from respective supplier elements;

and, (iii) perform the processing tasks to convert incoming feedstocks to outgoing products. Depending on the properties of incoming feedstocks and the specification of outgoing products, the exact tasks that a unit element performs and the type of services it requires can vary time-to-time.

- *Header Element:* The role of a header element is to transfer one or more materials or services within a segment of the process network. To satisfy this role, it executes the following functions: (i) identify the configuration of the process routes through which the materials or services are to be transferred; (ii) identify the requirements for property change for the materials or services being transferred (*e.g.*, heat, cool, pressurise, depressurise them); (iii) develop and implement a procedure(s) to switch the process routes from their current configuration to required target configuration; and, (iv) transfer materials or services in a controlled manner by interacting with respective unit or header elements.

- *Service Element:* The role of a service element is to distribute one or more services to its customer elements. To satisfy this role, it executes the following functions: (i) identify the nature of service demands from customer elements and decide the service supplies available for those demands; (ii) determine an optimal, or when necessary an emergent but sub-optimal, distribution of services while taking into account the priorities of service demands; and, (iii) distribute the services in a controlled manner via interacting with respective customer elements.[2]

- *Product Element:* The role of a product element is to represent the requirements of a production order and to ensure that these are met. To satisfy this role, it executes the following functions: (i) identify the processing tasks to be executed; (ii) map these tasks onto production capabilities of unit and header elements available in the plant; and, (iii) engage with unit and header elements to allocate these tasks (see next chapter for more details on how this mapping and allocation of tasks is carried out). Unlike previous distributed architectures and as discussed in Section 3.2, the product elements do not directly coordinate the operations of unit or header elements or the flow of materials or services in the network; this is done by unit, header and service elements themselves via direct interactions.

Fig. 4.3 depicts an UML diagram (Unified Modelling Language) of the data model and control functions of all four element types. The *association* relations, shown by solid lines, denote the presence of interactions between

---

[2] A service element may comprise its own internal production system to produce services. This process can be similarly represented via appropriate unit, header and service elements. When referred to the main system, the service element then also acts as a type of product element representing the composite demands of customer elements requesting its services.

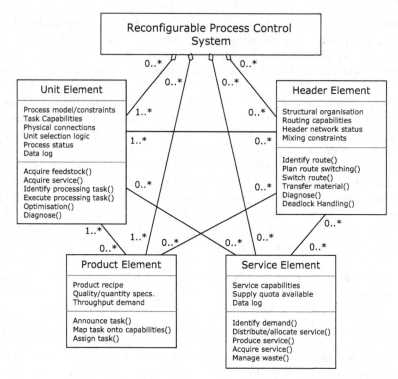

Fig. 4.3. Data model and control functions of process elements

process elements while the *aggregation* relations, shown by solid lines with diamond heads, denote the aggregation of process elements into an RPC system (Booch & Rumbaugh 2005). As the multiplicities in the figure suggest, a continuous process based on an RPC system must have at least one unit element and one product element in order to be able to produce a product. As the complexity of the process grows with more unit elements included and these elements sharing various services, the architecture requires adding further header and service elements.

4.2.3 Internal Structure of Process Elements

Internally, each process element is considered a self-contained control function comprising its local control module, a co-ordination module and the associated (optional) physical process part as shown in Fig. 4.4.

The internal design is derived initially from a decomposition of the multi-level control hierarchy as illustrated in Fig. 4.5. Each layer in the hierarchy is split along the physical dimension, followed by integrating vertically the localised blocks into control and coordination modules of process elements. The control module in Fig. 4.4 covers the execution functions (*i.e.,* basic

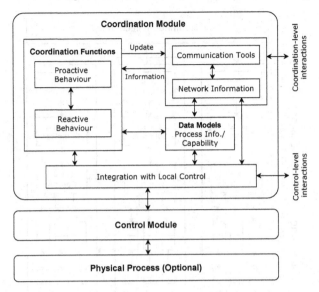

Fig. 4.4. Internal structure of process elements

control and some advanced control functions) and the coordination module the decision functions (*i.e.*, advanced control and levels above in Fig. 4.5). In addition to this, new components are included within coordination module to define the data models (process structure, capability *etc.*), the coordination functions (proactive and reactive behaviour) and the communication means to interact with other elements. Each process element thus receives the ability to plan, optimise and control its operations plus that of the relevant global system by coordinating with other elements.

4.2.4 Physical Connections Between Process Elements

Because of the manner in which basic element types are identified (*i.e.*, based on physical structure of the process instead of functional hierarchy), the process elements remain connected via material and service streams at process level. Fig. 4.6 depicts the five categories of such connections.

- *Material flow between unit elements:* This flow leads to production of the end-products. The flow may occur on a forward path (from raw-materials to end-products) or on a recycle path (from recovery units back to upstream units or intermediate storage).
- *Service flow to unit elements:* This flow may be required for the execution of processing tasks of unit elements (*e.g.*, supply of steam or cooling water for a reaction task).
- *Service flow to header elements:* This flow supports the transfer of materials or allows changing their properties, *e.g.*, heat or cool them.

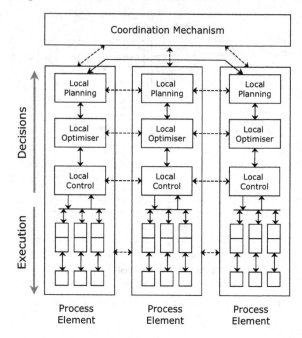

Fig. 4.5. Decomposition of multi-level control hierarchy

- *Exchange of services between unit and header elements:* This flow refers to recovered services from unit elements to be reused in the process (*e.g.*, heat released by exothermic reactor can be used to heat other materials)
- *Exchange of services between header elements:* This flow refers to recovered services from header elements to be reused in the process.

Note that the product elements do not have a physical presence and are not shown in Fig. 4.6 or described here. Their role is to provide unit and header elements with the product recipe information and they do so at the coordination level.

4.3 Migrating to Process Elements

The identification and design of process elements, being of fully distributed form, can be a radical change to the design practices currently in use in the industry. In order that these new concepts can be experimented – at least partially – using the commercial off-the-shelf tools available in DCS and PLC architectures, we can consider a migration approach based on an incremental decomposition of the control hierarchy. Previously, such an approach was also suggested by Chirn & McFarlane (2001) in the context of a discrete manufacturing architecture.

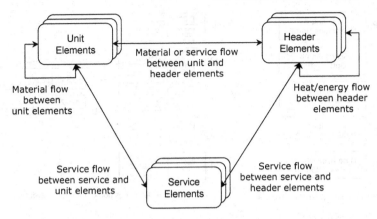

Fig. 4.6. Physical connections between unit, header and service elements

As a first step to migration, it is proposed that only the individual levels in hierarchy are decomposed as shown in Fig. 4.7(b); Fig. 4.7(a) shows the existing structure. These individual levels can still be implemented separately as in a conventional system using commercial tools of today. The individual levels may require developing a separate problem solving mechanism to enable them solve the control problems in a distributed form.[3] Next, one or more levels in the hierarchy in this distributed form should be integrated vertically (*e.g.*, optimisation and advanced control) so as to distribute more of higher level decisions down to lower levels (Fig. 4.7(c)). Finally, all levels in the hierarchy should be integrated vertically (Fig. 4.7(d)) such that the decisions requiring coordination are made by the coordination modules of elements and the execution of the outcomes of decisions is carried out by the control modules. It is envisaged that these vertically integrated design can be *packaged* together with respective physical process parts of process elements and supplied as stand-alone components to be plugged into an RPC system.

4.4 An Illustrative Example

We now apply the proposed architecture to a polymer process example shown in Fig. 4.8. The purpose of the example is to illustrate the selection and basic functions of process elements within an industrial process.

4.4.1 Description of the Process

The example process comprises two independent production lines, each comprising two polymerisation reactors. The process starts with reaction between

---

[3] In Chapter 6 a method to achieve this form of distribution is developed for a simplified control problem relating to the optimisation or advanced control levels.

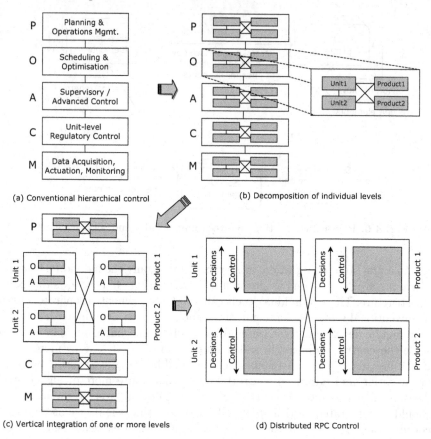

Fig. 4.7. Migration approach for developing internal designs of process elements

two main raw-materials (monomer and demineralized water) in the presence of other ingredients. This results in a slurry form of the end-product containing un-reacted raw-materials. The un-reacted raw-materials are separated in the flash vessels and stripper columns while the purified product is dehydrated and dried in the centrifuge and drier units before sent for storage. The recovered monomer is compressed and cooled for reuse again in the main reaction. The whole process operate in a batch-semicontinuous type, *i.e.*, the reactors operate in a batch mode while the other units in a semicontinuous mode.

The polymer end-product is supplied in a solid-grain form and is used in the manufacture of plastic products, *e.g.*, roof sheets, tanks, films and bottles. Depending on the type of application, the polymer grade that is used may differ in various chemical properties. The process considered here is capable of producing five grades (called grades A to E), each having further sub-grades. While all five grades use the same sequence of unit operations, the processing tasks used for each may vary in terms of the reaction conditions, separation

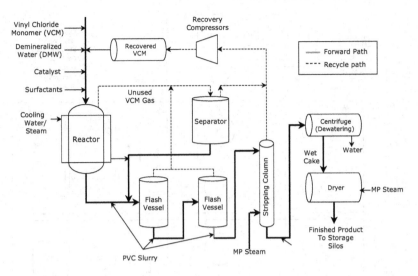

Fig. 4.8. Polymer process example

temperatures and the ratio and quality of recycled monomer to be allowed to be mixed with the fresh monomer. For instance, in a 'film' grade product only fresh monomer is allowed with stringent control of reaction conditions to achieve the desired quality of final product.

The process units consume various services for their processing tasks. The reaction occurs at a temperature between $50 - 90°C$, hence the feedstock entering the reactor is first heated to bring it to this temperature before it can enter the reactor. The reaction itself is exothermic and releases heat. This is removed via circulating cooling water (atmospheric temperature) and chilled water ($4°C$) in jacket and baffles of the reactor. The stripper columns use pressurised steam to purify the slurry and remove un-reacted monomer. The purified slurry thus contains extra heat which is conserved by heating the feed entering the reactor. The drier units similarly use pressurised steam to dry the purified slurry into a solid form.

Fig. 4.9 shows the layout of the process considered here. The layout offers a level of physical flexibility in terms of each line comprising certain parallel equipment that can be interconnected in various combinations. Each line can also be configured to produce a different polymer grade independently of the other line. Within each line, the reactors operate in a batch mode, hence individual reactor can be set up to produce a specific sub-grade without causing a significant mix up. The three stripper columns are shared between both lines such that only two columns are in operation while the third is being cleaned and regenerated. A similar facility exist to interconnect the centrifuge and drier units between lines, but this is not normally used unless necessary.

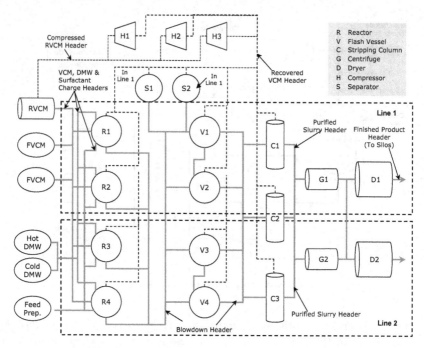

Fig. 4.9. Layout of the selected polymer plant

4.4.2 Identification of Process Elements

Based on the description in Section 4.2.1, the selection and assignment of process elements in the above process can be done as follows:

- *Unit Elements:* All process units, namely reactors, flash vessels, stripper columns, centrifuges, driers, and compressors are represented by individual unit elements. Each unit element possesses its local decisions that it regulates on its own. The reactor element, for instance, decides the yield of reaction (percentage of monomer converted to polymer) while the flash vessels and stripper columns decide the recycle flow of monomer. Note that these local decisions of unit elements interact physically due to their recycle connections. There often exists trade-offs. For example, the cost of separation and purification can be reduced if the reaction yield is increased, however this also means slow reaction times and hence reduced overall throughput. An optimum selection of local conditions is thus necessary to achieve global production goal.

- *Header Elements:* All 'switchable' piping networks in the process are represented by individual header elements. The identification is done based on materials or services being transferred, *e.g.*, monomer header, slurry header (or so-called blowdown header), purified product header, cooling water header, *etc.* Although not shown in Fig. 4.9, the header elements also contain transfer equipments and heat exchangers where necessary.

- *Service Elements:* The supplier plants for all services (*e.g.*, cooling water, chilled water, pressurised steam) are represented by different service elements. An effective distribution of services remains crucial to plant operations. For example, the stripper columns must receive a minimum supply of steam at a pressure above certain value in order to produce an on-grade product. If the available supplies drop suddenly, then shedding of steam supply to other unit elements, *e.g.*, to driers, may be necessary with a simultaneous reduction in production throughput by unit elements.

- *Product Elements:* Each customer order for a separate sub-grade is represented by a product element. Multiple product elements may exist, although only two would be produced at a time as only two lines are available. The customer orders may span a number of campaigns, each campaign comprising multiple batches. A detailed schedule of recipe information, quality, quantity, *etc.* may be supplied as part of the definition of product elements but not all parameters may be needed depending on how the process of recipe management is managed between product and unit elements (see the next chapter for details).

4.5 Comments on the DRPC Architecture

Having proposed the new architecture and illustrated its use, we use the final section of this chapter to reflect on the features and properties of the architecture.

4.5.1 Comparison with Conventional Process Control

The proposed architecture differs from conventional approaches in that it is modular and distributed. The plant-wide control is decomposed into control modules of process elements. The network-level response of process operation emerges via direct interactions between these elements. To this end, it is first shown that the proposed architecture is compatible with conventional control systems. Apart from that, its distributed nature offers certain additional benefits in terms of the two main attributes: improved product/process diversity and modifiability, where it supersedes conventional architectures.

Compatibility with Conventional Control

The four process elements are sufficient to cover all control functions commonly performed by a conventional control system. In a fully distributed case, the vertical hierarchy of control is decomposed into local control modules of elements. In an ideal case, the boundaries between planning, scheduling, optimisation and control levels are also blurred. The process elements solve the plantwide control problem at different level of abstraction depending on the nature of operating conditions and the disturbances arising, *i.e.,* operate in a long-term planning or proactive mode when there are no disturbances and in a short-term reactive mode when frequent disturbances are likely to arise. In the course of migrating to this fully distributed case, one can still consider an intermediate hierarchical form in which the elements first solve a planning problem in a distributed form and using its solution decide the set-points for lower-level scheduling or optimisation problems. By restructuring the control algorithms in this way one can develop and implement the same control functionality of a conventional system but now in a distributed way. This argument hence proves the sufficiency.

The functions of unit, header and service elements also relate directly to the physical decomposition of a continuous process into its constituent elements, *i.e.,* process units, piping networks and service suppliers. The product recipes of product elements specify the requirements of customer orders to bring together the physical elements to derive a process scheme. This is illustrated in Fig. 4.10. The presence of all four types of process elements is thus essential in a typical medium-to-large size plant to produce an end-product. The four types are thus also necessary to cover all control functions of a conventional control system.

Improved Product and Process Diversity

By using the notion of a product element, the architecture separates the procedural aspects of equipment control from the technical aspects of product recipes. This separation is important since it allows for the modification of both aspects in run-time. For instance, in case of frequent disturbances, it might be more sensible to consider an alternative product recipe or processing scheme than to reschedule the entire operation. Moreover, as discussed in the next chapter in detail, this integration is delayed until the stage where the actual production of a specific order is required, hence the most recent status of conditions on the plant can be taken into account.

Although the dynamic integration of recipe information can equally be incorporated in a conventional approach, the distributed nature of the architecture provides a sensible framework to implement it for two reasons. Firstly, the procedural control of process units is distributed, hence the equipment control can be easily modified. Secondly, the actual integration of product recipe occurs in a bottom-up manner by localised assignment of tasks to unit elements. As a result, emerging changes or disturbances can be managed in a graceful manner compared to conventional systems where this requires rescheduling the parts of or the full operation.

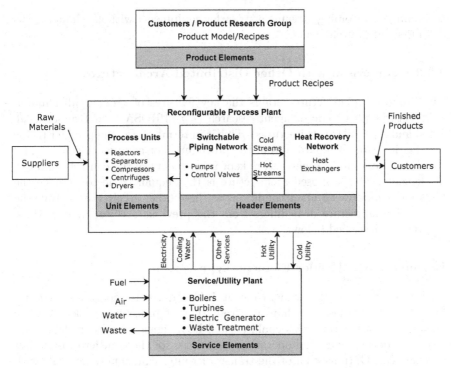

Fig. 4.10. Basic process elements cover all functions in a process plant

Easy Modifiability

The decoupling in local control of unit, header and service elements in the architecture can be expected to improve the modifiability of the control system. For example, one can easily replace a unit element with an equivalent another unit element provided their interfaces to other elements are same or compatible. The decoupling allows developing generic, multipurpose design of elements that can be standardised across a range of processes and re-used with little design and engineering effort as it has been the case for various package systems used in the industry, *e.g.*, industrial refrigeration.

Also, the use of header elements in the architecture introduces a further level of decoupling between unit elements. Unlike conventional control designs where each unit controller is pre-defined with exactly which other unit controllers it is connected with, in the proposed model the unit elements acquire this information on physical connectivity via header elements. The modifiability of the control system is hence enhanced in two respects: (a) it allows unit elements can be added or removed without changing the structure of their connections to other unit elements, *i.e.*, they only need to be defined with the header elements they are connected with, and (b) the level of flexibility supported in the design of header elements can be changed as necessary (*e.g.*, by adding extra piping streams or transfer equipment) without changing the definitions of unit elements.

4.5.2 Comparison with Other Distributed Architectures

The proposed architecture retains the key features of previous distributed architectures in holonic or agent research (*e.g.*, PROSA (van Brussel *et al.* 1998) and its similar architectures). This is in terms of: (a) the product recipe information is kept separate from the procedures for equipment control, and (b) the architectural properties are kept independent of the control strategies of elements. The proposed architecture is thus equally able to manage the foreseen or unforeseen plant conditions as these other architectures. But, the proposed architecture also differs from the previous architectures in three respects as described below.

Introduction of Header Element Type

A new element type, the header element, is introduced to separate the control functions of transport mechanisms from that of processing tasks. A more advanced function than this could as well be assigned to header elements – that is to derive sequential operating procedures for the transfer of materials and services. Optimisation of the transport routes can equally be dealt with by header elements. Since these functions are implemented independently of the unit or service elements, a new form of decoupling is achieved that should help improve the modifiability of the control system.

Introduction of Service Element Type

A new element type of service element is also introduced to address the absence of a separate mechanism in other architectures to support the distribution of services. Although such functions are equally important in discrete manufacturing, they play an absolutely vital role in the timely and correct operations of process plants. Note that the service elements do not directly perform any processing tasks, hence could not, and should not, be represented by unit elements.

Specification of Interaction Behaviour of Process Elements

While the interaction behaviour of process elements is discussed at length in the next chapter, it suffices here to say that the role of product elements in the proposed architecture is different to that in the other architectures. The product elements interact with unit elements to map the product recipes onto production capabilities in the plant. Unlike PROSA or related architectures though, the product elements do not manage the logistics of materials or services in the network nor do they define the operating conditions of these other elements. Such decisions are made by unit or header elements themselves once they are assigned with their tasks in the production.

4.6 Summary

In this chapter we have proposed a distributed architecture to support the reconfigurable process control. We next go onto examine how these elements in the architecture interact to allow the process to operate.

An Interaction Model for Reconfigurable Process Control

5.1 Introduction

The previous chapter developed a reconfigurable architecture for process control comprising four basic types of process elements. As discussed these elements must interact in order to exchange information and make collective decisions required for a complete plant-wide control. In this chapter we develop a model to support such interactions so as to describe how elements carry out information exchanges in order to produce the required end-products. The proposed model is a direct and necessary supplement to the reconfigurable control architecture developed in the previous chapter.

5.1.1 Overview

We again start by defining the term *interaction* and the meaning of an *interaction model* in the context of a distributed control architecture.

The term interaction – defined in a dictionary term as *action or influence of persons or things on each other* (Oxford English Dictionary, 2005) – has different meanings when referred to distributed architectures.These include: (i) *communication* (a pre-determined, passive message-passing protocol), (ii) *collaboration* (communication with dynamic selection of messages from an application-specific library), and (iii) *coordination* (collaboration with an ability to reason about and synthesise messages dynamically) (Wooldridge 2002). While the first two forms are used vividly in day-to-day life, it is the third form that has found convincing use in distributed decision-making applications such as in multi-agent systems, manufacturing control, supply chain management, *etc.* Coordination is also the meaning used in our interaction model, however, the current chapter only focusses on the collaboration aspects as defining the structure of information exchanged between process elements. The strategies for reasoning about local decisions are addressed in the next chapter.

The previous work in holonic or agent-based research has generally used contracting or its extended variant as the means for defining the message structure for inter-element interactions between product and resource holons. As discussed in Section 3.2, contracting alone cannot be applied to continuous processes because of the various restrictions such as the tight and finite interconnections between process units and the continuous flow of materials. These constraints can make the interactions significantly complex if the responsibility of managing unit, header or service elements is passed solely to product elements. Instead, we seek an interaction approach that allows these three latter elements to coordinate their operations directly among themselves while also interacting with product elements.

We seek inspiration from the research on information and life-cycle management in supply chain and virtual enterprise fields (Camarinha-Matos *et al.* 2003, Strader *et al.* 1998). The proposed model is then framed around developing an interaction method that process elements can use to implement the entire reconfiguration process shown earlier in Fig. 3.1 in a distributed manner, *i.e.*, how they identify the need for reconfiguration; define the new configuration; reorganise the process schemes; and, deliver the order requirements. The interaction model essentially builds upon a key concept analogous to market transactions between companies: a customer process element (*e.g.*, a unit element) that needs to acquire a material or service for its task can buy it from a supplier process element that can supply it, *i.e.*, the interactions between unit, header and service elements in the DRPC architecture are modelled as forms of supplier-customer type transactions between companies in a supply chain.

5.1.2 Requirements for RPC Interaction Model

The interaction model (including the coordination strategy to be developed in the next chapter) is aimed to deliver the dynamical aspects of reconfigurability requirements, in particular, product and process diversity, responsiveness and fault-tolerance. In addition, it should retain the modularity of the distributed architecture, *i.e.*, it should not impose any centralised information constraints that can disturb the modifiability of the RPC system. The key focus of this chapter is then to define the method for initial integration and its subsequent refinement of product recipe information with production capabilities in the plant so as to address these requirements in a distributed manner.

This chapter is structured as follows. The next section specifies the structure of process elements' interactions to implement the process of reconfiguration. The key important phases of this model are also explored with details about how the low-level message-passing protocols are organised for exchange of materials and services between elements. Section 5.3 then explains the interaction model by applying it to a small process example. Section 5.4 finally discusses the key features of the model by comparing it against the above requirements and the previous conventional and distributed approaches.

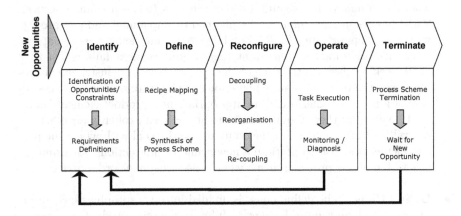

Fig. 5.1. Distributed reconfiguration process

5.2 Specification of the Interactions Between Process Elements

Fig. 5.1 depicts the distributed reconfiguration process to define the interactions of process elements in the DRPC architecture. The figure is developed from an earlier Fig. 3.1, with an additional 'Terminate' phase is now included and the overall activities of reconfiguration are split into five key phases.

We now describe each of these phases in a brief detail. To provide consistency in the discussions, we use the terms *product recipe, processing task*, and *process scheme* in the description below. A *product recipe* in the sense of ISA-S88 standard (ANSI/ISA 1995) refers to the minimum set of information that uniquely defines the manufacturing requirements for a specific product, *i.e.*, it identifies raw-materials, their relative quantities and the required processing, but without referring to particular equipment. A *processing task* in a product recipe refers to a unit operation (*e.g.*, reaction) that converts its incoming feedstocks to outgoing products. The recipe may define relative quantities of materials (and services if used) but not the exact values as they depend on the throughput of final products. A product recipe is then defined as a sequence of processing tasks that converts the main raw-materials to final products. A *process scheme* refers to the physical and control configuration of a segment of process that meets the requirements of product recipe for a specific order. Multiple process schemes may exist in the same process for different product orders sharing some unit, header or service elements common between them. Not all unit, header or service elements may be part of a process scheme though, *i.e.*, they can be idle at times and await for a new product order to arrive that is relevant to them.

- **Identify Phase:** The reconfiguration starts with identifying an opportunity for production (when a new production order arrive) or deciding whether to adapt the ongoing process schemes (when a change occurs). Such changes in particular can be planned or unplanned and can provide with a new opportunity (*e.g.*, availability of a unit element, raw-material or service) or impose a constraint (*e.g.*, a unit element fails or becomes bottleneck). Subsequently, each of the opportunities or constraints are refined into the detailed requirements for reconfiguration of the process. In case of a new customer order, an appropriate new product element is created. (Note that these requirements are not defined explicitly anywhere or centrally within an element; they only define the goal with which the process elements set out to initiate a new round of interactions, for example, to develop a new process scheme when a new customer order arrives.)

- **Define Phase:** The define phase is divided into two sub-phases: (i) *recipe mapping*, to map the product recipe information onto production capabilities available in the plant, and (ii) *synthesis*, to derive a specific process scheme from the potential choices created by recipe mapping.

  - *Recipe Mapping:* In the recipe mapping phase, the product elements associated with customer orders interact with the unit and header elements in the plant to assign (or refine already assigned) processing tasks in product recipes with the production capabilities available in the plant. The interactions lead to a number of tentative process schemes which could be used in the production. Not all tentative schemes may be feasible though because the selection of specific unit or header elements or their operational settings are not defined yet.

  - *Synthesis:* Next, in the synthesis phase, the unit elements involved in the tentative schemes interact among themselves as well as with the header and service elements to refine these tentative schemes into a single scheme that can be implemented for the production. The elements use a global production goal, such as production cost, to arrive at the solution. It is possible that some elements may be already engaged with other process schemes. Such elements interact among associated elements in those schemes to identify how should they reconfigure their operations. Eventually, all concerned elements agree on the structure of the new process scheme including various network parameters (process routes, material and service flow rates, *etc.*) and the operating settings of participating elements (reaction temperatures, heat exchanger duty *etc.*).

    As explained later in this section, the unit elements build the process scheme in an incremental, bottom-up manner by following demand-pull type interactions. In doing so, they also solve both a scheduling problem (defining the configuration of the scheme, *i.e.*, exact assignment of tasks to equipment) and an optimisation problem (defining the configuration of control structures and associated local settings).

- **Reconfigure Phase:** The procedure for reconfiguring the process scheme can now begin. The activities can be split conceptually into three sub-phases: decouple, reorganise and recouple, where any or all three of them may involve a physical and/or a control change. A good example of physical change can be a change in the process routes. The header elements involved in this change switch the routes from their current configuration to agreed target configuration. A systematic operating procedure may be required to meet the physical process constraints such as mixing hazards, cleaning-in-place, *etc.*

- **Operate Phase:** As the process scheme is being established, the flow of materials and services can also begin. The unit elements along the route start executing their processing tasks in a coordinated manner so as to convert the incoming feedstocks to their products. The coordination of all activities leading up to the start-up and subsequent on-load operations is achieved by unit elements themselves. The aim is to maintain the process at agreed set-points during synthesis phase.

  During continued operations, the plant conditions may change, *e.g.*, a unit element fails or the customer demand changes. The process elements affected by the change respond to it in a graceful manner. The elements situated next to the point of change attempt to absorb it to the level possible within their local capacity. If this is not achievable, the residual change is propagated further in the process up to a point where it can be fully absorbed. The elements affected along the route adapt their operations as appropriate. If the change or disturbance is small in magnitude or is not likely to last long, then the elements may prefer to operate in this mode for a required period. Only if it is large in magnitude or if the resulting performance is not acceptable, the elements should re-enter into a new round of interactions to reconfigure the process scheme starting with the identify phase.

- **Terminate Phase:** The process scheme is finally dissolved once the throughput requirements for the order are met or if a major failure occurs (such as a reactor element fails) which requires terminating the order altogether. In either case, the process elements involved in the scheme either join other process schemes or idle themselves and wait for a future order to arrive.

In what follows, we elaborate on the two sub-phases – recipe mapping and synthesis – of the define phase, as they are the most critical in terms of how the elements define the structure of a new process scheme. A specification of the underlying interactions described below would depend on the final application and is not developed in great detail here.

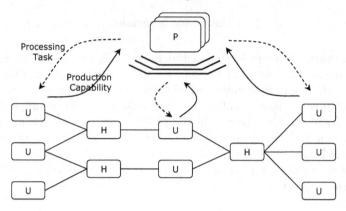

Fig. 5.2. Interactions between product and unit elements during the recipe mapping

5.2.1 Recipe Mapping Phase

Fig. 5.2 illustrates the structure of the interactions during recipe mapping phase. To illustrate the different possibilities, we consider two alternative ways in which this mapping can be carried out: (i) product-centric approach and (ii) unit-centric approach.

Product-Centric Approach

In a product-centric approach, the product elements take the leading role in assigning the processing tasks to unit elements. Each product element announces each of its tasks in the product recipe to all unit elements existing in the plant. The unit elements capable of undertaking the tasks reply back. The replies may contain primary information about the nature of unit operation, *e.g.*, 'reaction type 1', that the unit elements can perform. Based on the replies, the product element identifies the unit element(s) that best suite the currently announced task and assigns the task to them. It also assigns appropriate materials and services as well as other processing requirements for that task (*e.g.*, product quality) to selected unit elements. The interactions repeat in this manner until all tasks in the product recipe are assigned to appropriate unit elements. Note that the same task may be assigned to more than one unit elements but not all may be able to undertake the task, because the information on the connectivity of these unit elements or their local settings is not defined as yet. The interactions thus result in one or more tentative process schemes that are refined into a single scheme in the synthesis phase.

Unit-Centric Approach

The centralised role of product elements in a product-centric mechanism can become bottleneck if they have to match a large number of tasks or the same tasks more frequently. This latter scenario can arise if the process is required to produce the same end-products more frequently and/or in a highly reconfigurable manner (*e.g.*, in case of many polymer plants).

Instead of product elements, the responsibility of recipe mapping can be distributed among unit elements themselves. The unit elements can now be defined with additional details about the specific processing tasks they can perform and the materials and services they need to acquire to execute these tasks. For instance, a distillation column can be specified with two specific distillation tasks $X \Rightarrow \{Y, Z\}$ and $L \Rightarrow \{M, N\}$ (where X, Y, Z, L, M and N are the materials). The product elements can still be supplied with some form of product recipe or parts of it if the customer order requires only certain processing tasks to be used in making the product.

Using recipe-specific details, the unit elements can be asked to identify which task(s) they can use to produce a specific material. The recipe mapping activity then proceeds as a backward search starting from the unit elements that can produce the final product. These unit elements first identify the tasks they can use to produce the product and the incoming materials they require from other unit elements in the upstream. The unit elements may check with product elements whether their selected tasks are not restricted in the recipe (if supplied). The interactions repeat from the upstream unit elements until the unit elements requiring the main raw-materials are reached. The interactions thus result in one or more tentative process schemes which could be refined into a single scheme during the synthesis phase. Since the synthesis phase is also carried out by unit elements themselves (together with header and service elements) it is possible that the recipe mapping and synthesis can proceed together.

Discussion

It can be seen that both approaches to recipe mapping have their benefits and disadvantages – the product-centric approach may require less time to set up initially while the unit-centric approach may provide increased freedom to unit elements to choose or alter their tasks. The unit-centric approach requires unit elements to be defined with additional recipe information on processing tasks they can perform. This is not a requirement for product-centric case. The initial effort required to set up a unit-centric approach may thus be higher. However, the unit-centric approach also allows unit elements to select or alter their processing tasks that best match with the changing plant conditions. This is instead of those specified by the product elements in a product-centric approach. The unit-centric approach should thus be able to better utilise the plant facilities than the product-centric approach.

We must note that the inclusion of recipe-specific information in unit elements in the unit-centric approach does not change or violate the basic assumption that this product information should be kept separate from production capabilities as considered in ISA-S88 standard (ANSI/ISA 1995) or the previous distributed research. What the approach suggests is to derive this information in a bottom-up manner by collecting together the tasks of unit elements via their direct interactions into a single scheme. As mentioned earlier in this section, an approach of this nature may be useful when the same products are produced more frequently or the same tasks are reused in different products.

5.2.2 Synthesis Phase

Having identified the processing tasks, the unit elements in the tentative process schemes interact among themselves and with respective header and service elements to identify the structure of a specific process scheme that can be used for production. The interactions are considered to follow a backward-search pattern based on demand-pull in which the unit elements, starting from the terminal stage of the process, attempt to incrementally allocate their material demands to unit elements situated upstream and also the service demands to appropriate service elements. Fig. 5.2 outlines the nature of these interactions.

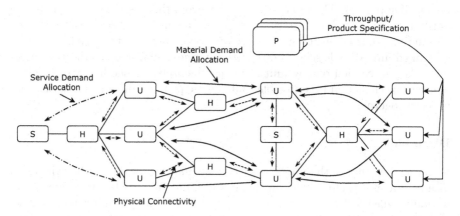

Fig. 5.3. Interactions between unit, header and service elements during the synthesis phase

Internal Design of Process Elements

To model the interactions in detail, we borrow an analogy from the concept of so-called *transaction* between companies in a market or a supply chain in that 'a process element (which can be a unit or a header element) that requires a material or service for execution of its processing task can *buy* this from any other process element or elements which can supply it', *i.e.*, the exchange of a material or service between process elements can be modelled as a form of contract between two or more different parties in a market or supply chain.

Using this analogy, we can impose a structure on the internal design of process elements, in particular on their coordination modules in Fig. 4.4. A process element which requires access to a material or service for its task can be represented as a customer and a process element that supplies a material or service can be represented as a supplier. Fig. 5.4 depicts this structure. This suggests that each unit element can be modelled as: (a) the supplier of its outgoing products and services (*e.g.*, heat released from exothermic reaction) and (b) the customer of its incoming feedstocks and services. Similarly, each header element can be modelled as a supplier or customer for its supply or use of services and each service element purely as the supplier of its services.

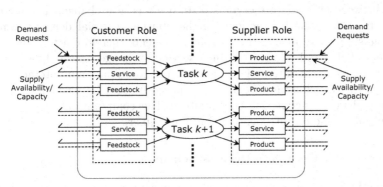

Fig. 5.4. Internal design of process elements based on supplier/customer roles

Interaction Protocol

Based on the supplier-customer roles, the interaction protocols for allocation of material or service demands between two or more process elements can be defined to follow a specific time line:

- Step 1: the customer element announces a demand request for supply of relevant material or service;
- Step 2: the supplier elements which can supply the material or service respond to these requests;
- Step 3: the customer and supplier elements agree on the allocation of material or service demand in terms of respective process parameters, *e.g.*, process flow rates, temperature, pressure.

If the exchange of material or service is to occur via a header element, then those header elements also get involved in the interactions so as to agree on the process routes though which the transfer should occur as well as the requirements for transforming the physical state of the material or service being transferred, *e.g.*, heat or cool them.

Figs. 5.5 and 5.6 show the interaction protocols for material and service demand allocation between a customer process element (denoted as theCustomer) and one or more supplier process elements (denoted as theSupplier). The exchange occurs via intermediate header elements (denoted as theHeader). In the case of material exchange, the customer element must be a unit element, while the supplier element can be other unit element or an external supplier when it is the main raw-material. Similarly, for the service exchange, the customer element can be a unit or header element while the supplier can be a unit, header or service element depending on where and how the service is supplied.

Note that the protocols in Figs. 5.5 and 5.6 differ in the way the interactions between elements are organised. In a material exchange, the customer elements initiate the interactions for distributing material demands among possible supplier elements; the customer elements therefore act as the coordinators of demand allocations. In a service exchange, again the customer elements initiate the interactions, however the coordination of interactions in terms of the distribution of service is achieved by supplier elements, *i.e.*, the supplier elements act as the coordinators. The computational methods for implementing these protocols therefore must differ.

Synthesis of a Complete Process Scheme

The synthesis of a complete process scheme from tentative process schemes identified during recipe mapping occurs via a sequence of nested material and service exchanges between unit, header and service elements. Fig. 5.7 on page 83 depicts an overview of these interactions between unit elements. All unit elements therein are shown both as customers and suppliers of their feedstocks and products as well as services.

The round of interactions starts from unit elements in the last stage. These unit elements initiate the protocol for material and service allocation in Figs. 5.5 and 5.6 by announcing the demand requests for their feedstocks and services. For material demands, the interactions proceed in backward direction. Not all unit elements in the upstream in tentative schemes which can supply feedstocks may respond because there may exist constraints such as limited connectivity or physical capacity limits; the connectivity information is supplied by the header elements. The unit elements which can meet the supply requirements further initiate a new set of material and service allocation protocols to source their feedstocks from unit elements in further upstream and services from appropriate service elements. The interactions thus repeat until unit elements in the first stage of the process are reached that can acquire their feedstocks from raw-material suppliers. At this stage, starting from the first stage, all concerned unit elements return back with their supply proposals (including the availability and capacity details) to respective customer

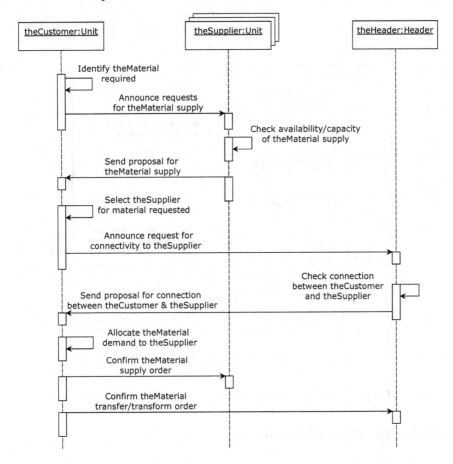

Fig. 5.5. Interaction protocol for material demand allocation

elements. The supply proposals thus flow in forward direction towards the terminal stage. From the responses, each unit element selects which supplier elements are appropriate, and how much material demand it should allocate to them. If necessary, this nested sequence of interactions for material and service allocations repeats until all participating elements settle on respective parameters for material and service demands as well as the process routes through which the transfers should occur. The process scheme thus developed is then reconfigured in the next 'reconfigure' phase which is not described here in detail.

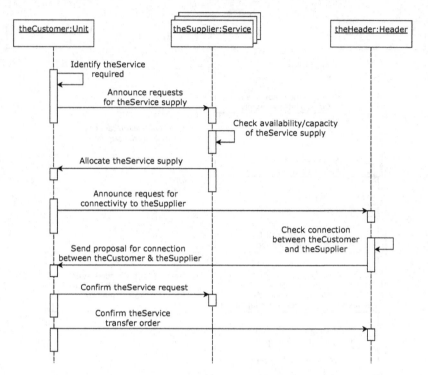

Fig. 5.6. Interaction protocol for service demand allocation

5.3 An Illustrative Example

We next consider a simple process example to illustrate the nature interactions between process elements in the DRPC interaction model. The example illustrates a production start-up scenario where a specified product is to be produced at a given throughput rate. A further detailed example depicting the general production control scenarios such as multiple products is considered in Chapter 7.

5.3.1 Process Description

We consider a process where a product A is produced using the product recipe shown in Fig. 5.8. Each rectangle in the figure represents a material and each oblong a processing task. Each task is associated with at least one outgoing and one incoming material, whereas each material with at least one task. The recipe is of a *non-linear* nature, *i.e.*, there exist two different tasks T3 and T4 that both can produce material D and hence be involved in producing A. Fig. 5.9 depicts the layout of the physical process comprising a set of unit, header and service elements.

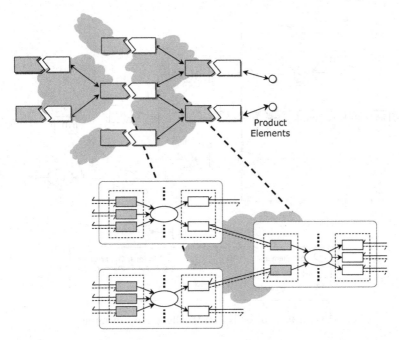

Fig. 5.7. Interactions between unit elements in the synthesis of a complete process scheme

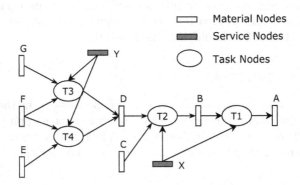

Fig. 5.8. Product recipe for illustrative example

5.3.2 Application of the DRPC Interaction Model

As per DRPC architecture, a product element PR is associated with product A. Each unit, header and service element is represented via a unit, header and service elements with symbols respectively as U, H and S. The (external) supplier elements for raw-materials are represented via prefix R. The process contains alternative process schemes through which materials A can be produced. These include U1, U2 in combination with any of the suppliers

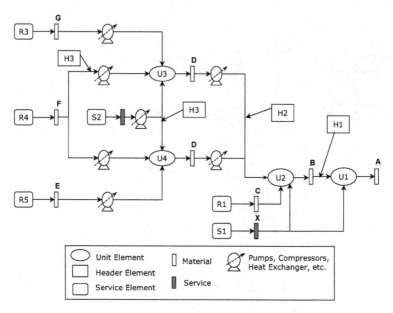

Fig. 5.9. Process layout for illustrative example

of D, *e.g.*, {U3}, {U4}, or {U3,U4}. The following sub-sections illustrate the sequence of interactions between product and unit elements using product-centric and unit-centric approach for recipe mapping (the header and service elements are omitted for simplifying the illustration).

Product-Centric Approach

Fig. 5.10 illustrates an animated sequence of interactions in the proposed model when a product-centric approach is used for recipe mapping. The individual steps therein can be described briefly as follows (the terms in brackets show the phase in interaction model to which the step corresponds to).

Step 0 : (*identify*) A new product element PR is created;

Step 1 : (*recipe mapping*) PR starts with announcing task T1 in recipe;

Step 2 : (*recipe mapping*) Since U1 can only execute T1, it replies back;

Step 3 : (*recipe mapping*) PR assigns T1 to U1;

Step 4 : (*recipe mapping*) Interactions repeat until all tasks in recipe are assigned;

Step 5 : (synthesis) U1 and then U2 send material requests to their supplier unit elements as identified in Step 4. The supplier elements return their proposals;

Step 6 : (*synthesis*) U1 and then U2 allocate their material demands to supplier elements. The configuration of process scheme is thus fixed;

Step 7 : (*reconfigure & operate*) The process scheme is reconfigured (as appropriate). Material and service flows are established;

Step 8 : (*terminate*) Process scheme is terminated when order requirements are met. PR is removed.

Unit-Centric Approach

Fig. 5.11 illustrates the sequence of interactions when a unit-centric approach is used for recipe mapping. Again the individual steps therein can be described briefly as follows.

Step 0 : (*identify*) A new product element PR is created;

Step 1 : (*recipe mapping*) PR announces its order requirement for producing product A;

Step 2 : (*recipe mapping*) U1 can produce A through task T1. It confirms with PR that T1 is allowed in product recipe;

Step 3 : (*recipe mapping*) U1 announces material request for B; U2 can supply B. It replies its interest;

Step 4 : (*recipe mapping*) U2 confirms its task T2 with PR and also extends the scheme to U3-U4;

Step 5 : (*recipe mapping*) U3 and U4 similarly confirm their tasks and extend the scheme;

Step 6 : (*synthesis*) U3-U4 and then U2 return supply proposals to their customer elements. U1 and then U2 then allocate their demands;

Step 7 : (*reconfigure & operate*) The process scheme is reconfigured (as appropriate). Material and service flows are established;

Step 8 : (*terminate*) Process scheme is terminated when order requirements are met. PR is removed.

Discussion

As can be seen from Figs. 5.10 and 5.11, the role of product element PR in the interactions is limited to that of assigning tasks (product-centric approach)

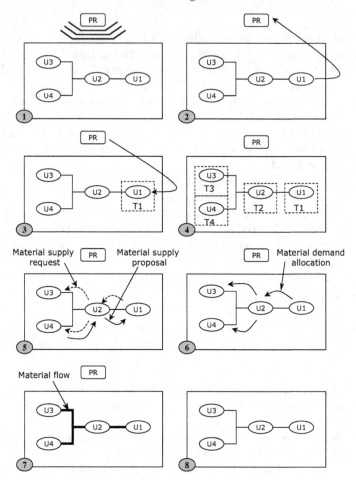

Fig. 5.10. Illustration of interaction model using product-centric approach

or confirming that the tasks selected by unit elements are allowed in the recipe (unit-centric approach). The actual synthesis of process scheme in terms of deciding the process parameters and local settings is carried out by unit elements themselves (together with header and service elements). As discussed in detail in the next section, this distinction forms a key difference in the proposed model compared to earlier interaction models in holonic or agent research.

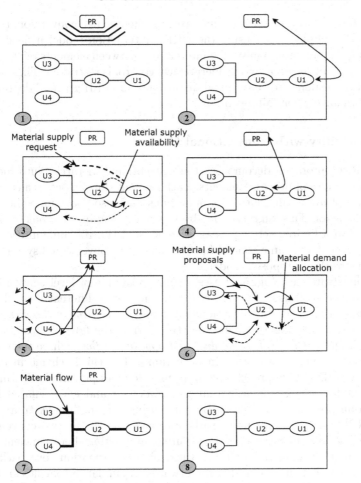

Fig. 5.11. Illustration of interaction model using unit-centric approach

5.4 Comments on the DRPC Interaction Model

To conclude this chapter, we comment on potential differences between the proposed interaction model and its analogs within existing conventional and alternative distributed approaches.

5.4.1 Comparison with Conventional Process Control

Conventional control is based on hierarchical information and control flow. As discussed in Section 2.3, the production in a hierarchical system is driven by a higher-level, long-term plan derived based on customer order forecasts (Williams 1989). Below, we first show that the interaction model described in this chapter is compatible to this conventional information flow in that all control functions and interfaces that are implemented in a conventional system can also be implemented using the DRPC interaction model if need be. In addition, the bottom-up nature of interactions between elements offers several new benefits in the areas where hierarchical control is restricted. These are: (i) bottom-up response to change and disturbance, and (ii) graceful degradation of performance when failures occur.

Compatibility with Conventional Control

The DRPC approach decomposes the conventional hierarchy into localised control modules of process elements. Each element thus possesses a capability to plan, optimise and control its operations and also coordinate them with other elements. This suggests that by restricting the interactions of process elements to a limited set of process schemes and control configurations, one can implement the same control functionality of a conventional system using the framework of proposed model.

To illustrate this, Fig. 5.12 uses a simple example of level control in two series-connected tanks. The figure also includes two different control structures used frequently in conventional systems: (a) *control in the direction of flow*, where the product demand directly controls the flow of incoming raw-material, and (b) *control in the direction opposite of flow*, where the demand is propagated via level control in both tanks. The third scheme in the figure shows a DRPC approach operating in a demand-pull mode. In this case the variable pairings for each tank are combined and encapsulated into a general-purpose control module. By configuring this module as appropriate, the DRPC model can be made to behave as either of the two conventional schemes, because in either case the nature of interactions between elements in DRPC model remains the same *i.e.,* the demand information flows backwards and the variations in material flows forwards. A similar argument can be extended to other control levels to interpret the information flow at those levels in a demand-pull form. This indicates the compatibility of DRPC model to conventional control.

Bottom-up Response to Change and Disturbances

The example in the previous section briefly demonstrated the manner in which the process elements interact to provide a response to arrival of a new product order. The response emerges via bottom-up (*i.e.,* element-to-element) interactions. This behaviour is not predefined in the interaction model. A more detailed example illustrating these issues is given in Chapter 7. The model is thus able to deal with different scenarios or a combination thereof without an explicit definition of global response for each case.

(a) Control in the direction of material flow

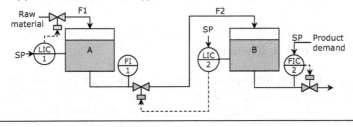

(b) Control in the direction opposite of material flow (On-demand scheme)

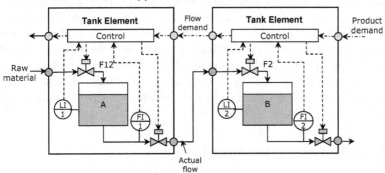

(c) Distributed RPC Scheme

Fig. 5.12. Comparison of conventional and DRPC control structures

Graceful Degradation of Performance in Case of Failures

The interaction model provides a definitive structure and guideline on how should elements exchange information. To enhance the predictability of operations the model guarantees that the interactions are flexible but also binding. A unit element, for example, should not simply de-commit from supplying its products to downstream unit elements in case if its local process becomes bottleneck or one of its supplier unit elements fails. Rather, it seeks an alternative supplier element for the same feedstock. Only under the circumstance where the disturbance is sufficiently large (in magnitude or of long-term nature), it opts to propagate the disturbance further or terminate its processing tasks.

The interaction model thus ensures that performance degrades gracefully until a point is reached where the disturbance can be fully absorbed.

5.4.2 Comparison with Other Distributed Interaction Models

As noted in the introduction, the principle of contracting has been the basis of research in most previous holonic or agent research (Gou *et al.* 1998, van Brussel *et al.* 1998, Chirn & McFarlane 2001). The proposed model extends contracting or market programming approaches by using a virtual enterprise based model so as to address the constraint of physical connections between process units. The model thus differs from the previous research in various ways.

The Role of Product Elements

In the proposed model, the product elements interact with unit and header elements to map the processing tasks onto production capabilities. Unlike other architectures though (*e.g.*, PROSA (van Brussel *et al.* 1998) or its related architectures), the product elements do not manage the logistics of materials or services in the network nor do they define the operating conditions of unit, header or service elements. Such decisions are made by these latter elements themselves once the processing tasks are assigned. This modification hence avoids the complexity of coordination if the product elements are allocated with this responsibility. Additionally, as it was shown with unit-centric approach, the distribution of recipe mapping provides unit elements with an increased freedom to select local tasks that best match with the current plant status.

Network Behaviour of Process Elements

As can be seen, the virtual enterprise paradigm provides an effective approach to contracting or market approaches in dealing with the physical constraints. The unit, header and service elements, for instance, are now made able to interact with: (a) other such elements on process connections, and (b) product elements on product-recipe related issues. Product elements instead behave purely as information servers monitoring the adherence to product recipes. Moreover, as mentioned in Section 4.5, the whole network operation is decoupled by the use of header elements, that similar to transporters in supply chains, can be made flexible as necessary (by adding extra piping streams or transfer equipment equipment) irrespective of other elements requiring their use.

The proposed model also operates on a demand-driven basis, *i.e.*, the unit elements build processing schemes (in the form of material and service exchange protocols) in backward direction starting from the end-products to raw-materials. Subsequently, any new demand changes imposed on the process are also propagated in the process in an incremental manner. This behaviour ensures that the production remains fitting to changing demands.

The use of backward search in building or extending process schemes (see Fig. 5.10 and Fig. 5.11) also guarantees that the resulting partial schemes are feasible, *i.e.*, physically implementable. This may not so with other architectures mentioned above where the interactions between product and resource elements generally follow a dispatching mode of task allocation, *i.e.*, the next task in sequence is only announced when the previous task is finished. This could possibly result in dead-locks and dead-ends where the product elements may find no further machine available to progress the partly finished parts. The proposed model avoids this scenario by ensuring that the unit elements (together with header and service elements) build a complete process scheme from raw-materials to end-products before the actual production commences.

Distribution of Information and Control Functionality

Unlike contracting, the proposed model also improves the distribution of information and control among elements as the product elements are no longer responsible for coordinating local operations. Hence, little or no production information (depending on the product or unit-centric approach used for recipe mapping) needs to be transferred to product elements. This feature can be of significant use when: (a) the number of product elements that can coexist in the process is large; (b) the product elements are designed and developed by teams situated remotely, or (c) multiple product elements share some of the materials or unit elements which can lead to deadlocks because the supplies of materials or these shared unit elements are likely to fail.

5.5 Summary

In this chapter we have proposed a distributed interaction model to support the run-time interactions of process elements in the control architecture. We next go onto examining the quantitative aspect of these interactions, *i.e.*, to define a distributed solution strategy for use of the process elements, in particular the unit elements, to identify their local operating settings during the define, reconfigure or operate phases.

6

A Distributed Algorithm for Reconfigurable Process Control

6.1 Introduction

We now develop a strategy to effectively coordinate the operation of the DRPC system. Predictably, this is in the form of a distributed algorithm for guiding and managing the distributed interactions of process elements to achieve suitable control settings. We focus in particular on the interactions between unit elements during synthesis phase of the reconfiguration process (Fig. 5.1) where the settings of each element are established. The proposed algorithm is used by unit elements to find control settings for their local and other network parameters once the physical layout of the specific process scheme to be used for production has been established.

6.1.1 Overview

Before beginning with the strategy development, we position the proposed approach in the context of existing distributed approaches.

As reviewed in Section 2.4, previous research on distributed approaches in control, both in manufacturing and other domains, has used distribution to solve large control problems by breaking them into multiple smaller problems referring to individual subsystems and then solving them either independently (Šiljak 1991) or via iterative coordination based on hierarchical (Mesarovic et al. 1970) or distributed (Bertsekas & Tsitsiklis 1989) techniques. Alternatively, when problems are already distributed, the question of problem solving is to coordinate the local solutions so as to ensure a global objective or constraint is satisfied. Problems of these nature arise in numerous large-scale domains as explored in Section 2.5 which just looks at few.

Previous research in distributed manufacturing paradigms of holonic and agent-based manufacturing control have taken a view of separating the control architecture from control algorithms to enhance reconfigurability of the architecture, i.e., the desire for reconfigurability of the architecture and software drives distribution rather than computational simplification. Bongaerts

(Bongaerts *et al.* 2000, Bongaerts 1998) in this sense used a mix of hierarchical hierarchical (fully distributed) control to switch between proactive (when conditions are planned) and reactive behaviour (when disturbances arise) of distributed holons in so-called PROSA architecture (van Brussel *et al.* 1998). More advanced work on distributed scheduling has used lagrangian decomposition (Liu & Sycara 1997, Gou *et al.* 1998) and market programming approaches (Váncza & Márkus 1998, Tharumarajah 2001, Shen, Wang & Hao 2006) to define the methods for resource allocation, *i.e.*, assignment of tasks to machines and scheduling of their start and end times.

In this chapter we use a distributed coordination technique of *nested decomposition*, studied previously for multi-stage optimisation problems (Ho & Manne 1974, O'Neill 1976, Wittrock 1985), to define the coordination strategy for process elements. Our rationale for using this approach is two-fold:

a. The process units in a continuous process remain tightly interconnected, therefore the coordination of their distributed settings should occur via direct interactions between them instead of achieved by separate product elements as in previous holonic and agent research. It is likely that if the latter approach is used the amount of coordination effort required could become excessive;
b. The approach of nested decomposition provides an economic interpretation that can be linked to the price and demand guided interactions between companies in a virtual enterprise, and so, to the use of this analogy in defining the protocol for material exchange in the interaction model.

The previous techniques in nested decomposition are not immediately applicable to process control problems though because they can only be linked to multi-stage process networks of series-connected form. Instead, we seek an extension which can be applied to process networks of arbitrary form.

6.1.2 Requirements for a Distributed Coordination Strategy

The coordination approach is expected to support the distributed nature of control architecture and interaction model in previous two chapters. This distribution was considered essential to promote a maximum level of reconfigurability in the design and interactions of process elements. Any numerical technique used as part of coordination must not disturb the reconfigurability, *i.e.*, the use of a centralised entity or constraint must be avoided. We also aim that the elemental sub-problems (as obtained after distribution) adhere to a common production objective which in this chapter is considered as the sum of all local costs. This is to ensure the distributed solution meets the optimality and global coherence of hierarchical control where possible. With regards to the four requirements in Fig. 2.7, the strategy is also expected to provide a level of responsiveness to variations in local problem formulations and other disturbances.

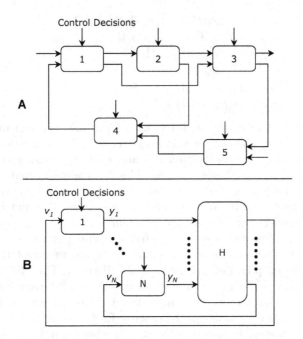

Fig. 6.1. Schematic of a large-Scale or complex system

This chapter is structured as follows. The next section introduces the formulation of distributed control problem to be used in this chapter as the basis of analysis. Section 6.3 characterises the technique behind overall solution strategy. Sections 6.4 to 6.6 then develop the distributed algorithm in a constructive manner. Illustrative examples and the potential future extensions of the approach are discussed in Sections 6.7 and 6.8.

6.2 Distributed Control Problem

We start by mathematically formalising the distributed control problem to be used in this chapter.

A chemical process can be seen as a large-scale system comprised of multiple subsystems or process units as shown in Fig. 6.1(a). In this network form, the outputs of each unit become the inputs to its downstream units and so on. By imposing an orderly input-output matrix H, this network form can be converted into a generic form shown in Fig. 6.1(b) where H now represents the network structure of the process, with each row in H referring to an input and each column to an output of associated process unit.

Assuming there are N units in the process, we then use the following model structure to define the dynamics of the whole process.

$$\dot{x}_i(t) = h_i(x_i(t), u_i(t), v_i(t), t) \quad i = 1, \ldots, N$$
$$y_i(t) = g_i(x_i(t), u_i(t), v_i(t), t)$$
$$v_i(t) = \sum_{\substack{j=1,\ldots,N \\ j \neq i}} H_{ij} y_j(t)$$
$$G_i(x_i(t), u_i(t), v_i(t), t) \in S_i$$
$$r(x(t), u(t), v(t), t) \in R$$

(6.1)

where $x_i \in X_i$ is the vector of states, $u_i \in U_i$ is the vector of manipulated variables, $y_i \in Y_i$ is the output vector and $v_i \in V_i$ is the interaction vector associated with unit i. The vectors x, u and v are the aggregate vectors of x_i, u_i and v_i, $i = 1, \ldots, N$ respectively. The constraints h_i and g_i represent the state and output equations, G_i are the local constraints, and r is the shared constraint coupling one or more process units. The matrix H_{ij} then aggregates the effects of all units $j \neq i$ on unit i. □

Assuming process units are connected via piping streams only, we can derive a specific formulation of matrix H. To do so, we adopt the so-called *P-Graph* model proposed by Friedler, Tarján, Huang & Fan (1992).

Omitting services (*e.g.*, steam, cooling water), in a P-Graph form the process is represented as a finite set of materials M being transformed by a finite set of process units O available in the process. Each process unit i in this sense is written as a material tuple $(\mathrm{mat}_i^{in}, \mathrm{mat}_i^{out})$ where mat_i^{in} and $\mathrm{mat}_i^{out} \in M$ are the sets of incoming and outgoing materials. If $\wp(M)$ is the set of all possible subsets of M, then we get the following two relationships:

$$O \subset \wp(M) \times \wp(M) \quad O \neq \emptyset,$$

and

$$M = \left\{ \bigcup_{i \in O} \mathrm{mat}_i^{in} \right\} \bigcup \left\{ \bigcup_{i \in O} \mathrm{mat}_i^{out} \right\}$$

We can then use the P-Graph (M, O) to represent the network structure of the process as a directed, bipartite graph comprising material nodes (as elements of M) and unit nodes (as elements of O). Fig. 6.2 shows an example P-Graph comprising three unit nodes, where M and O can be written as $M = \{A, B, C, D, E, F\}$ and $O = \{\mathsf{U1}, \mathsf{U2}, \mathsf{U3}\}$.

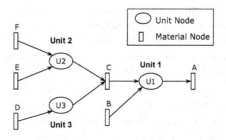

Fig. 6.2. An example of P-Graph model

In what follows, we consider a specific interpretation of vectors v_i, u_i, x_i and y_i in Eq. 6.1 to relate the model equations with demand-pull type interactions between unit elements in the interaction model.

- *Interactions v_i:* The interactions v_i are taken as the flow rate demands for outgoing materials mat_i^{out} of unit i, *i.e.*, the flow-rate demands act as a form of disturbances that the unit cannot control on its own but are set by the units in its downstream. Other types of interactions, *e.g.*, due to pressure and temperature variables, are omitted here.
- *Manipulated Inputs u_i:* The manipulated inputs u_i are divided into three types: $u_{i,in}$, $u_{i,util}$ and $u_{i,loc}$. The $u_{i,in}$ are set as the input flow-rates for materials mat_i^{in}, $u_{i,util}$ as the input flow-rates of utilities, and $u_{i,loc}$ as other local variables (*e.g.*, agitator speed of a reactor) associated with unit i. $u_{ij,in}$ as the jth element of $u_{i,in}$ then refers to unit i's input demand to unit j in its upstream.
- *States x_i:* The state variables x_i refer to various local properties of unit i, such as level, volume or material concentrations.
- *Outputs y_i:* The outputs y_i are taken as input demands $u_{i,in}$, *i.e.*, y_i for materials mat_i^{in} of unit i.

Using the above assignment of variables and P-graph model (M, O), we get at the following equivalent form of Eq. 6.1 where P-graph (M, O) is now used to replace matrix H.

$$\dot{x}_i(t) = h_i(x_i(t), u_i(t), v_i(t), t) \quad i = 1, \ldots, N$$
$$y_i(t) = \{u_{ij,in}(t)\} \quad \text{for all } j \in \mathbf{S}_i^q, q \in \mathrm{mat}_i^{in}$$
$$v_i(t) = \left\{ \sum_{j \in \mathbf{M}_i^d} u_{ji,in}(t) \right\} \quad \text{for all } d \in \mathrm{mat}_i^{out} \tag{6.2}$$
$$G_i(x_i(t), u_i(t), v_i(t), t) \in S_i$$
$$r(x(t), u(t), v(t), t) \in R$$

where, in reference to P-Graph (M, O), $\mathbf{M}_i^d$ and $\mathbf{S}_i^q$ are:

$\mathbf{M}_i^d$: indices of units $j \in [1, \ldots, N]$ connected to unit i
    through a material stream $d \in \mathrm{mat}_i^{out}$
$\mathbf{S}_i^q$: indices of units $j \in [1, \ldots, N]$ connected to unit i
    through a material stream $q \in \mathrm{mat}_i^{in}$. $\square$

In this chapter, to simplify the discussion and limit the amount of mathematical rigor involved, we restrict ourselves to a linear, steady-state form of Eq. 6.2 as follows.

$$0 = A_i x_i + B_i u_i - E_i v_i$$
$$y_i = \{u_{ij,in}\} \quad \text{for all } j \in \mathbf{S}_i^q, q \in \mathrm{mat}_i^{in}$$
$$v_i = \left\{ \sum_{j \in \mathbf{M}_i^d} u_{ji,in} \right\} \quad \text{for all } d \in \mathrm{mat}_i^{out} \tag{6.3}$$
$$x_i \in X_i,$$
$$u_i \in U_i$$

where x_i, u_i, v_i and y_i now all refer to their steady-state values. Matrices A_i and B_i in Eq. 6.3 are assumed to be of appropriate dimensions, while C_i, D_i, E_i and H_{ij} are assumed to possess a special form as discussed later in this section. Note that we have omitted the shared constraint $r(x, u, v) \in R$ in Eq. 6.2.

Fig. 6.3 depicts the example from Fig. 6.2 with the dynamics of individual units and their interactions. Note that process units are connected through relationship $v_i = \left\{ \sum_{j \in M_i^d} u_{ji,in} \right\}$ for all $d \in \text{mat}_i^{out}$. For units 2 and 3 the set M_i^d refers to unit 1 as the only customer unit for material C.

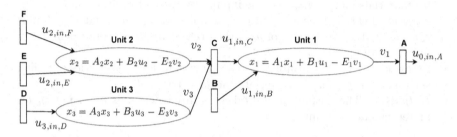

Fig. 6.3. Local unit dynamics for P-Graph in Fig. 6.2

In what follows we make a further assumption that the interaction vector v_i for all units $i = 1, \ldots, N$ is of the dimension equal to number of materials in mat_i^{out}, i.e., each element in v_i refers to a demand for a material in mat_i^{out}. If unit i supplies the same material to multiple customer units, then the total sum of all demands is used as the value for respective element in v_i.

With this assumption, the formulation of matrices C_i, D_i and E_i in Eq. 6.3 can be simplified. In particular, all entries in E_i are set to 0 except where a 1 appears when an element in v_i is connected to an element in state vector x_i via the state equation. The matrix C_i similarly becomes a zero matrix with all entries 0, while D_i becomes a matrix with all entries 0 except where a 1 appears when any element of $u_{i,in}$ is connected to an element of y_i. Note that each element in $u_{i,in}$ in this form is connected to only one element in y_i.

Using the revised form of E_i we can also simplify Eq. 6.3 by separating the rows in A_i and B_i which do not contain any element of v_i into separate local matrices $A_{i,loc}$ and $B_{i,loc}$. The constraints involving $A_{i,loc}$ and $B_{i,loc}$ then become the local constraints of unit i. For the sake of simplicity, we separate these constraints from remaining constraints and assume that A_i and B_i now only refer to those rows that correspond to an element of v_i. For simplicity, we also assume that each row in A_i and B_i is associated with only one element in v_i, i.e., the number of rows in A_i and B_i equal the length of v_i, which, in turn, equals the number of outgoing material streams $d \in \text{mat}_i^{out}$. We can then eliminate E_i from Eq. 6.3 altogether.

We now define the distributed control problem used in this chapter. Formally, the problem assigned to each unit $i = 1, \ldots, N$ is to find an optimal, steady-state deviation in its variables x_i and u_i from a nominal operating point $\bar{x}_i$ and $\bar{u}_i$ for a given demand deviation of v_i from the nominal demand $\bar{v}_i$. The nominal point for all three vectors may refer to a target set point supplied by the higher-level optimiser.

Problem 6.1 (Distributed Control Problem).

$$\underset{x_i, u_i}{\text{minimise}} \sum_{i=1}^{N} f_i(x_i, u_i) \quad i = 1, \ldots, N$$

$$\text{s.t.} \quad A_i x_i + B_i u_i = v_i,$$

$$A_{i,loc} x_i + B_{i,loc} u_i = 0,$$

$$y_i = \{u_{ij,in}\}, \quad \text{for all } j \in \mathbf{S}_i^q, q \in \text{mat}_i^{in} \qquad (6.4)$$

$$v_i = \left\{ \sum_{j \in \mathbf{M}_i^d} u_{ji,in} \right\}, \quad \text{for all } d \in \text{mat}_i^{out},$$

$$x_i \in X_i, u_i \in U_i \qquad \square$$

The objective function $f_i(\cdot)$ is assumed to be strictly convex jointly on its constituent variables x_i and u_i. For simplicity, we assume that f_i is linear-quadratic, *i.e.*, $x_i^T Q_i x_i + u_i^T R_i u_i + c_i^T [x_i^T, u_i^T]^T$ where $[]^T$ represents the transpose operator. This together with the affine nature of the constraint equations guarantees that the dual problem of Prob. 6.1 is differentiable (Rockafellar 1970).

In summary, the distributed control problem to be solved for the overall process is to minimise the joint total cost (as the sum of individual costs) of all units in the P-Graph subject to a constraint that all material flow interactions are satisfied between units. The solution of this problem then defines the material flow-rates $u_{i,in}$ in the network.

Note that although the local costs $f_i(x_i, u_i)$ of all units $i = 1, \ldots, N$ are separable, the individual sub-problems are not, because the constraints $v_i = \{\sum_{j \in \mathbf{M}_i^d} u_{ji,in}\}, \forall d \in \text{mat}_i^{out}$ link them. As a result the overall problem cannot simply be decomposed into sub-problems and solved independently. A distributed approach to solving Prob. 6.1 must be able to coordinate these linking constraints via distributed interactions.

Prob. 6.1 is general enough to be applied to a process network of any arbitrary nature. However, in this work, we limit ourselves to processes of *acyclic* nature only (*i.e.*, processes that do not contain material or energy recycles) and having no by-products. Recycles or by-products play an important role in modern process plants, however the developments made in this chapter cannot support such process forms at present.

We observe that this and other assumptions made in this section regarding problem formulation and network structure help us to simplify the solution strategy described next. These can be relaxed as appropriate by generalising the approach discussed here.

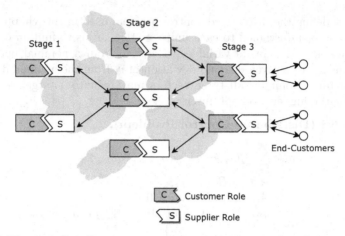

Fig. 6.4. Supplier-customer relationships between unit elements

6.3 Distributed Coordination Approach

Having defined the problem formulation, we now develop the distributed co-ordination approach used in this chapter for solving Prob. 6.1. We use the concept of so-called *nested decomposition* from optimisation and operations research literature (Ho & Manne 1974, O'Neill 1976, Wittrock 1985) to develop the approach. In simple terms, it refers to solving a multi-stage optimisation problem (*e.g.*, in a staircase type linear program) by solving multiple, smaller coordination problems, each associated with a linking constraint connecting two successive stages.

We split the development of the proposed approach into three main steps:

   i. *Problem decomposition:* Develop a method for decomposing the overall process network into multiple two-stage problems;
  ii. *Solution of two-stage problems:* Develop a general method for solving customer-supplier coordination (the two-stage problems);
 iii. *Solution of multi-stage problems:* Develop an algorithm for solving the two-stage problems in a nested sequence.

Before describing each of these in detail (Sections 6.4-6.6), below we provide in this section an outline of the overall approach.

Fig. 6.4 repeated here from Chapter 5 depicts the form of supplier-customer type relationships between unit elements in the interaction model – the unit elements act as the suppliers for their outgoing products and customers for their incoming feedstocks. In order to formalise this relationship mathematically, we consider a further analogy of price-demand type interactions between companies in a supply chain, or specifically a virtual enterprise, when operating under a make-to-order environment.

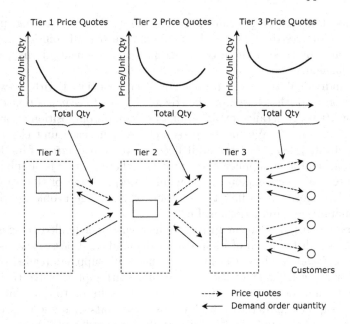

Fig. 6.5. Price-demand oriented interactions between companies in a supply chain

In a supply chain, as shown in Fig. 6.5, companies share price quotes and demand quantities to make decisions on which suppliers to select and how much demand is allocated to each supplier. The customer companies along each tier request potential supplier companies upstream for price quotes for expected demand quantities. The price quotes then flow forwards. The customer companies use these price quotes to select suppliers and allocate demands. The interactions thus remain bidirectional and may repeat until all companies settle on a price-demand contract, at which point the operations can commence.

When viewed in an analogous manner, we can formalise the interactions between unit elements by attaching a price quote to a supply proposal as the variation in local cost of the supplier unit element for a unit change in the product flow rate. The interactions would then proceed as follows.

Starting from the terminal stage in the process, each unit element in its role as a customer together with all unit elements acting as its suppliers attempt to form a two-stage control problem of the form of Prob. 6.1 involving only these unit elements. The objective is to distribute the customer's feedstock demands among suppliers in a manner that the total cost of suppliers and of the customer is minimised. The same principle is then applied to these supplier elements who attempt to distribute their demands by forming appropriate two-stage control problems involving unit elements further upstream. The process is repeated until unit elements in the first stage of the process are reached.

At this stage, we obtain multiple two-stage problems involving unit elements from successive stages of the process. The overall solution approach then operates by solving these two-stage problems sequentially in a nested, iterative fashion.

Each individual two-stage problem is itself associated with a two-level coordination algorithm based on the so-called *primal decomposition* concept (Geoffrion 1970). A detailed explanation of primal decomposition approach is given in Appendix A. Within the two-stage problem, each unit element has two roles. In its role as a customer, it becomes the *coordinator* of its demand distribution, and as a supplier a *sub-problem* where its role is supplying the price quotes. We next exploit an economic interpretation of the primal decomposition algorithm to link the solution of two-stage problems with the price-demand type interactions in Fig. 6.5.

At first, each unit element in its role as a customer (called the customer unit) selects a distribution of its feedstock demands and passes that as *coordination variables* to all unit elements acting as its suppliers (called supplier units). For a given demand, the supplier units attempt to solve their local problems to find the optimum supply costs and a solution to other local variables. The supplier units return back to customer units their supply proposals comprising: (a) the supply cost for the specified demand, and (b) the *marginal cost* as an indication of the variation in supply cost for a unit change in the demand flow rate. In the language of primal decomposition, this supply proposal refers to so-called *optimality cut* that is included in customer unit's local problem at the next iteration. Subsequently, at each iteration, the customer unit adjusts its demand, taking into account the previous supply proposals, such that the total cost of all suppliers plus its own is minimised.

The overall interactions across the process then operate in a chain whereby all these two-stage problems are solved iteratively in a nested, iterative sequence. Starting from terminal stage, each individual customer unit allocates its feedstock demands to its supplier units. These supplier units then further propagate the demands upstream until the first stage is reached. Next, starting from the first stage, the supplier units return their supply proposals (as optimality cuts) back to customer units downstream. The customer units in each stage then include the new supply proposals to solve their local problems and alter the demands allocated. This multi-level chain of interactions thus repeat, first backwards and then forwards, until all interim demand flow rates converge to an optimum value.

The solution strategy described above supports the reconfigurability of the interaction model in following ways. Firstly, the above solution approach operates in a completely distributed form, *i.e.*, the overall problem is solved by direct interactions between unit elements themselves without referral to a centralised coordinator. Secondly, because all unit element problems are made independent via distributed interactions, the approach allows adding or removing unit elements from problem formulation without having to reformulate the dynamics model in Prob. 6.1 which otherwise would be necessary in a centralised implementation. Thirdly, under certain conditions it can be shown that the solution obtained for interim flow rates converges to the same optimal solution as that obtained by solving a centralised problem. Finally, if there are bounds on supply capacities of the supplier units that can be modelled in the formulation of customer problems, then it is possible to restrict the interim demands of customer units to be within these bounds to ensure feasibility. The above solution process, hence, can be interrupted at any stage in the sequence to use a suboptimal but immediately usable solution. This might be desirable, for example, in a constantly changing environment. The proposed approach thus retains distributed character of the interaction model whilst also maintaining optimality of the solution.

In the next three sections we now address the three steps of the proposed distributed coordination approach in detail.

6.4 Problem Decomposition

In the first step to solving Prob. 6.1, the overall problem is decomposed into a set of two-stage problems, each referring to a network junction between two or more process streams. We refer to such a junction as a *Junction Block*. Each junction block is thus a two-stage process consisting all unit elements and process streams associated with that junction.

Fig. 6.6 shows the four different types of junction blocks that can be found in any acyclic process network. The MIXER and SPLITTER blocks represent the junctions associated with process units such as mixer, splitter or a piping header where multiple material streams of identical properties are mixed together or a single stream split into multiple such streams. The MULTIFEED and MULTIPROD represent the junctions associated with process units such as feed preparation, reactor, distillation column *etc.*, where material streams of non-identical properties are merged together or are produced as outcomes of the processing task. Note that a more complex junction with multiple incoming and outgoing streams can be always represented by superimposing MIXER or MULTIFEED blocks on top of SPLITTER or MULTIPROD blocks.

Fig. 6.7 shows an example of problem decomposition in which an arbitrary process network is decomposed into its constituent junction blocks. The junction blocks are interconnected via process units common between them.

6.5 Solution of Two-Stage Problems

In the next step to solving Prob. 6.1, each junction block, being a two-stage process, is associated with a two-level coordination algorithm based on primal decomposition concept of the type introduced in Section 6.3. This algorithm is used to solve the associated mini control problem of junction block involving local problems of unit elements in both stages in a distributed manner. The

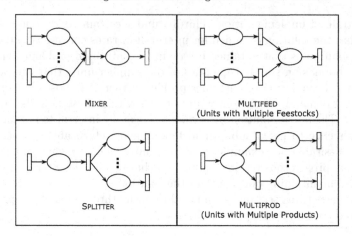

Fig. 6.6. Four types of 'junction blocks' in an acyclic process network

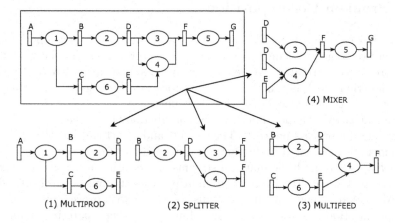

Fig. 6.7. Decomposition of a process network into junction blocks

unit elements in the second stage become the *coordinators* or so-called *master* problems and those in the first stage as so-called *sub-problems*. In economic terms, the first and second stage elements also act as suppliers and customers of materials through which they are connected in the P-Graph.

Since all four junction blocks in Fig. 6.6 have different network structures, *i.e., joins* in MIXER and MULTIFEED blocks and *forks* in SPLITTER and MULTIPROD blocks, they need different coordination techniques. The problems for MIXER and MULTIFEED blocks involve a single coordinator and can be solved based on primal decomposition technique as described in Appendix A. However, for SPLITTER and MULTIPROD blocks, this is not directly possible as they involve multiple customers and hence multiple coordinators. Therefore, we develop a variant of primal decomposition using the techniques from parametric programming field (Fiacco 1983) to accommodate this variation.

In what follows, we develop a single algorithm that can be applied to all four types of junction blocks with a facility to tailor the algorithm for individual type of block. To develop the necessary algorithm, we consider solving three problems of increasing complexity:

i. Two-Units problem involving two, series-connected units (Section 6.5.1)
ii. Two-Units problem with an uncontrolled parameter (Section 6.5.2)
iii. Multi-unit problem, so-called *superset block* problem, which enables the solution to all four junction blocks (Section 6.5.3).

6.5.1 Solution of Two-Units Problem

We consider first a process network comprising two unit elements connected in series, the so-called STAIRCASE block. Based on Prob. 6.1, the associated control problem for this block can be written as:

Problem 6.2 (Two-Units Problem).

$$\begin{aligned} \underset{\substack{x_1,u_1 \\ x_2,u_2}}{\text{minimise}} \quad & f_1(x_1,u_1) + f_2(x_2,u_2) \\ \text{s.t.} \quad & A_1x_1 + B_1u_1 = v_1 \\ & A_2x_2 + B_2u_2 = v_2 \\ & x_1 \in X_1, u_1 \in U_1 \\ & x_2 \in X_2, u_2 \in U_2 \end{aligned} \tag{6.5}$$

where $v_1 \equiv y_2 = u_{21,in}$ represents the demand from unit 2 to unit 1.□

In the proposed use of primal decomposition, unit 2 becomes the coordinator or master problem (denoted by SP_2), unit 1 as the only sub-problem (denoted by SP_1), and v_1 is the interaction variable linking them. The solution process then operates iteratively. For an initial value of $\hat{v}_1$, the problem SP_1 is solved first to find the optimal values of the value function α_1 and the Lagrange multiplier λ_1 for linking constraint $A_1x_1 + B_1u_1 = \hat{v}_1$. This information is passed as an optimality cut to the master problem SP_2. With including this new cut, the master problem SP_2 is solved to find a revised value of $\hat{v}_1$ that minimises the total cost of both problems. The process thus repeats between solving SP_1 and SP_2 until a form of convergence is achieved. Algorithm 6.1, based on the description in Section A.2, Appendix A, formally defines this solution procedure.

Fig. 6.8 outlines the information exchange between SP_1 and SP_2 problems. Note that ν_2 in SP_2 represents the supremum of piecewise-linear approximations of SP_1's optimal value function $\alpha_1(\hat{v}_1)$. This together with $f_2(x_2,u_2)$ equals the approximate total cost of both unit elements at any one iteration. Given that the optimality cuts do not overestimate $\alpha_1(\hat{v}_1)$, the optimal cost obtained from solving SP_2 provides a lower-bound on the total cost of both units. Since each new optimality cut should improve upon the linear approximation of $\alpha_1(\hat{v}_1)$, the lower-bound obtained should improve at each iteration until SP_2 converges to an optimal solution.

Algorithm 6.1 (Two-Units Problem)

Step 0: Initialise: Set $K := 1$. Assume an initial value of $x_2^{(0)} \in X_2$ and $u_2^{(0)} \in U_2$. Set $\hat{v}_1^{(0)} \equiv y_2^{(0)} = C_2 x_2^{(0)} + D_2 u_2^{(0)} = u_{21,in}^{(0)}$.

Step 1: Sub-problem $SP_1^{(K)}$: At any iteration K, fixing $\hat{v}_1^{(K-1)} \equiv y_2^{(K-1)}$, solve unit 1's problem $SP_1^{(K)}$ as:

$$SP_1^{(K)} \quad \begin{cases} \underset{x_1, u_1}{minimise}\ \alpha_1 \triangleq f_1(x_1, u_1) \\ s.t. \quad A_1 x_1 + B_1 u_1 = \hat{v}_1^{(K-1)} \\ \quad\quad x_1 \in X_1, u_1 \in U_1 \end{cases} \quad (6.6)$$

Set $z_1^{(K)} \equiv A_1 x_1^{(K)} + B_1 u_1^{(K)}$. Pass $\alpha_1^{(K)}$, $\lambda_1^{(K)}$ and $z_1^{(K)}$ to unit 2.

Step 2: Master Problem $SP_2^{(K)}$: Use $\alpha_1^{(K)}$, $\lambda_1^{(K)}$ and $z_1^{(K)}$ to construct a new optimality cut in unit 2's problem $SP_2^{(K)}$. Solve the resulting problem as:

$$SP_2^{(K)} \quad \begin{cases} \underset{x_2, u_2, \nu_2}{minimise}\ \alpha_2 \triangleq \nu_2 + f_2(x_2, u_2) \\ s.t. \quad \nu_2 \geq \alpha_1^{(k)} + \lambda_1^{(k)}(z_1^{(k)} - u_{21,in}), \quad k \in \mathbf{K} \\ \quad\quad A_2 x_2 + B_2 u_2 = \hat{v}_2^{(K-1)}, \\ \quad\quad x_2 \in X_2, u_2 \in U_2, \\ where \quad \mathbf{K} \triangleq set\ of\ iterations \end{cases} \quad (6.7)$$

Step 3: Terminate/Iterate: Terminate if the convergence criteria is satisfied, which is considered to be as

$$\left\| \left\{ x_1^{(K)}, u_1^{(K)}, x_2^{(K)}, u_2^{(K)} \right\} - \left\{ x_1^{(K-1)}, u_1^{(K-1)}, x_2^{(K-1)}, u_2^{(K-1)} \right\} \right\| \leq \epsilon. \quad (6.8)$$

Else, set $K := K + 1$ and return to step 1. $\square$

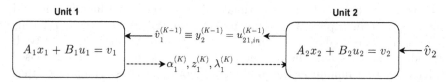

Fig. 6.8. Information exchange between units 1 and 2 sub-problems in Two-Units problem

6.5.2 Solution of Two-Units Parametric Problem

We consider next an extension of the Two-Units problem (Prob. 6.2) where we now assume that SP_1 contains an uncontrolled parameter vector θ which it cannot alter.

Problem 6.3 (Parametric Two-Units Problem).

$$\text{minimise}_{\substack{x_1, u_1 \\ x_2, u_2}} f_1(x_1, u_1) + f_2(x_2, u_2)$$

$$\begin{aligned}
\text{s.t.} \quad & A_1 x_1 + B_1 u_1 = v_1 \\
& A_{1,\theta} x_1 + B_{1,\theta} u_1 = \theta \\
& A_2 x_2 + B_2 u_2 = v_2 \\
& x_1 \in X_1, u_1 \in U_1 \\
& x_2 \in X_2, u_2 \in U_2
\end{aligned} \tag{6.9}$$

where $\theta \in \Theta$ is a vector of additional parameters local to unit 1. Assume that Θ is a convex, compact sub-set of $\mathbb{R}^{p_\theta}$ where p_θ denotes the dimension of θ. $\square$

If Algorithm 6.1 for Two-Units problem is applied to the above problem, then sub-problem $SP_{1,\theta}$ would have the following formulation at an iteration K.

$$SP_{1,\theta}^{(K)} \left\{ \begin{aligned}
& \text{minimise}_{x_1, u_1} \alpha_{1,\theta} \triangleq f_1(x_1, u_1) \\
& \text{s.t.} \quad A_1 x_1 + B_1 u_1 = \hat{v}_1^{(K-1)} \rightsquigarrow \lambda_1^{(K)} \\
& \qquad A_{1,\theta} x_1 + B_{1,\theta} u_1 = \theta^{(K)} \rightsquigarrow \lambda_{1,\theta}^{(K)} \\
& \qquad x_1 \in X_1, u_1 \in U_1
\end{aligned} \right. \tag{6.10}$$

where $\theta^{(K)}$ denotes the value of θ at iteration K.

Considering that the second constraint in $SP_{1,\theta}^{(K)}$ is a function of x_1, u_1 and θ, any change in θ from $\theta^{(K)}$ would change the feasible region of (x_1, u_1) and hence their optimal values for a given $\hat{v}_1^{(K-1)}$. Any such variation therefore would lead to a non-unique response from $SP_{1,\theta}$ to master problem $SP_2^{(K)}$ for a given $\hat{v}_1^{(K-1)}$. A significant variation in θ may also invalidate the optimality cuts passed to master problem in the previous iterations.

The situation can be recovered if the Algorithm 6.1 is restarted at every instance when θ changes. However, this is undesirable if θ is likely to change frequently (as in the case considered in the next sub-section). Instead, we propose a simple alteration to Algorithm 6.1 which interprets the change in θ and updates the previous optimality cuts passed to the master problem. The proposed technique is based on so-called *basic sensitivity theorem* from sensitivity analysis studies (Fiacco 1983) and is referred to as the *approximate optimality cut update technique*. See Section A.1 in Appendix A for a brief discussion of the relevant concepts from sensitivity analysis.

We first consider the following first-order approximation result based on basic sensitivity theorem (Theorem 3.2.2 in Fiacco 1983).

Lemma 6.4. *Consider the optimal solution* $\alpha_{1,\theta}^{(K)}$ *of the value function* $\alpha_{1,\theta}$ *in Eq. 6.10. Assume the constraint matrix* $\begin{bmatrix} A_1 & B_1 \\ A_{1,\theta} & B_{1,\theta} \end{bmatrix}$ *has a full row rank, i.e., the rows are linearly independent, then*

$$\frac{\partial \alpha_{1,\theta}^{(K)}}{\partial \theta} = -\lambda_{1,\theta}^{(K)} \tag{6.11}$$

where $\lambda_{1,\theta} \in \Lambda_{1,\theta} \subseteq \mathbb{R}_{+}^{p_{\lambda}}$ is the vector of Lagrange multipliers for the second constraint $A_{1,\theta}x_1 + B_{1,\theta}u_1 = \theta$, and $\lambda_{1,\theta}^{(K)}$ denotes its value for a given $\theta = \theta^{(K)}$. $\square$

The lemma suggests that for a change of $\Delta\theta$ in θ from $\theta^{(K)}$, the optimal value of the value function α_1 can be expected to change at least by $\Delta\alpha_{1,\theta} \equiv -\lambda_{1,\theta}^{(K)}$. Note that this is still a first-order approximation, and cannot be used to find the exact change in $\alpha_{1,\theta}^{(K)}$. However, even when $\Delta\theta$ is large, the following is true.

Lemma 6.5. *Consider the value function $\alpha_{1,\theta}$ in Eq. 6.10. If the function f_1 is convex in (x_1, u_1) and the sets X_1 and U_1 in Prob. 6.3 are convex, then the optimal value function $\alpha_{1,\theta}$ is convex on Θ. Furthermore, for a change in θ from any $\theta^{(j)}$ to $\theta^{(k)}, k \geq j, \theta^{(j)}, \theta^{(k)} \in \Theta$,*

$$\alpha_{1,\theta}^{(k)} \geq \alpha_{1,\theta}^{(j)} - \lambda_{1,\theta}^{(j)^{T}} (\theta^{(k)} - \theta^{(j)}), \tag{6.12}$$

where $\alpha_{1,\theta}^{(j)}$ and $\alpha_{1,\theta}^{(k)}$ are the optimal values of $\alpha_{1,\theta}$ obtained by solving Eq. 6.10 for value of θ being $\theta^{(j)}$ and $\theta^{(k)}$. In other words, the change in optimal value of $\alpha_{1,\theta}$ as obtained from solving Eq. 6.11 for a change in θ from $\theta^{(j)}$ to $\theta^{(k)}$ is always an underestimation of the optimal value of $\alpha_{1,\theta}^{(k)}$ that results by solving the sub-problem in Eq. 6.10 again at $\theta^{(k)}$.

Proof. The first part of the statement follows from standard convexity results in parametric nonlinear programming (see *e.g.*, Proposition 2.1 in Fiacco & Kyparisis 1986) while considering in addition that both constraints in Eq. 6.10 are linear in (x_1, u_1). The second part follows due to convexity of $\alpha_{1,\theta}^{(j)}$ in θ and noting that $\alpha_{1,\theta}^{(j)} - \lambda_{1,\theta}^{(j)^{T}} (\theta - \theta^{(j)})$ is a linear support to optimal $\alpha_{1,\theta}^{(j)}$ at $\theta = \theta^{(j)}$. $\square$

Fig. 6.9 illustrates the intention behind considering above lemma. The bold curve therein shows the variation in optimal value function $\alpha_{1,\theta}^{*}$ as a function of θ with $\hat{v}_1^{(K-1)}$ being constant, while the straight line shows the gradient to $\alpha_{1,\theta}^{*}$ at $\theta = \theta^{(j)}$. By using this gradient, we can obtain an approximate value of $\alpha_{1,\theta}^{*}$ for $\theta = \theta^{(k)}$ to update the optimality cuts in the master problem from previous iterations.

The modified procedure then operates exactly the same as Algorithm 6.1 except the following. At Step 1, the values of parameter θ and the Lagrange multiplier $\lambda_{1,\theta}$ for constraint $A_{1,\theta}x_1 + B_{1,\theta}u_1 = \theta$ from the previous iterations and the current iteration (respectively as $\theta^{(k)}$ and $\lambda_{1,\theta}^{(k)}, k \in \mathbf{K}$) are used to construct an update vector $\alpha_{1,updt}^{(k)}, k \in \mathbf{K}$, where $\alpha_{1,updt}^{(k)}$ is calculated as:

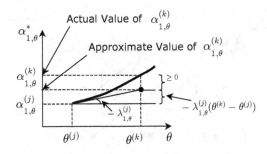

Fig. 6.9. Linear approximation of $\alpha_{1,\theta}^{*}$ as a function of variation in θ

$$\alpha_{1,updt}^{(k)} \triangleq -\lambda_{1,\theta}^{(k)^{T}}\left(\theta^{(K)} - \theta^{(k)}\right), \quad k \in \mathbf{K}. \tag{6.13}$$

At Step 2, $\alpha_{1,updt}^{(k)}, k \in \mathbf{K}$ is then used to update the optimality cuts in the master problem SP_2 before solving the revised SP_2 as:

$$SP_2^{(K)} \begin{cases} \underset{x_2,u_2,\nu_2}{\text{minimise }} \alpha_2 \triangleq \nu_2 + f_2(x_2, u_2) \\ \quad \text{s.t.} \quad \nu_2 \geq \alpha_{1,\theta}^{(k)} + \alpha_{1,updt}^{(k)} + \lambda_1^{(k)}(z_1^{(k)} - u_{21,in}), \quad k \in \mathbf{K} \\ \qquad\quad A_2 x_2 + B_2 u_2 = \hat{v}_2, \\ \qquad\quad x_2 \in X_2, u_2 \in U_2, \\ \quad \text{where} \quad \mathbf{K} \triangleq \text{set of iteration indices} \end{cases} \tag{6.14}$$

Compared to Algorithm 6.1, the above modification thus requires calculating $\alpha_{1,updt}^{(k)}, k \in \mathbf{K}$ as a reflection of the change in optimal value function $\alpha_{1,\theta}$ for a change in θ. The modified Algorithm 6.1 is not described here for brevity.

Note that in the above modification we still retain the same multipliers $\lambda_1^{(k)}, k \in \mathbf{K}$ in the master problem as before. In fact, the use of sensitivity analysis suggests that $\lambda_1^{(k)}$ also vary with a change in θ. Arguments similar to Lemma 6.4 can be used to obtain a first-order approximation of multipliers (Fiacco 1983). However, our numerical experience (see Section 6.7) indicates that using the same $\lambda_1^{(k)}$ do not lead to a significant problem considering that $-\lambda_1^{(k)}$ denote the sub-gradient of value function $\alpha_{1,\theta}^{(k)}$ for a unit change in $\hat{v}_1^{(k-1)}$. As a result, unless $\lambda_1^{(k)}$ changes widely, the hyperplane $\nu_2 \geq \alpha_{1,\theta}^{(k)} + \alpha_{1,updt}^{(k)} + \lambda_1^{(k)}(A_1 x_1^{(k)} + B_1 u_1^{(k)} - u_{21,in})$ still underestimates the value function $\alpha_{1,\theta}^{(k)}$ as required by the approximate cut update technique.

6.5.3 Solution of Superset Block Problem

The approximate cut update technique can be used to develop a solution algorithm for SPLITTER and MULTIPROD blocks. In particular, for each cus-

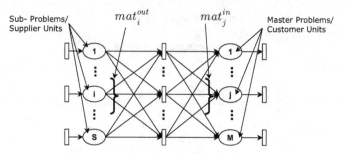

Fig. 6.10. Superset junction block

tomer element j in the second stage of either of these junction blocks, the demands $u_{mi,in}$ from all other customer elements $m \neq j$ can be treated as an uncontrollable parameter vector θ in the sub-problem of the supplier unit. The procedure in the modified Algorithm 6.1 can hence be repeated for all customer elements separately to coordinate the parametric effect of their demand changes onto the supplier element's problem.

Below we build upon this logic by developing a generic algorithm which can be applied to all four junction blocks, in particular the SPLITTER and MULTIPROD blocks. To do so, we consider a *superset* junction block as shown in Fig. 6.10 which captures within it all four types of blocks in Fig. 6.6, *i.e.*, the configuration of any block Fig. 6.6 can be obtained by selecting the appropriate edges and nodes in Fig. 6.10 while deleting the rest. Table 6.1 shows the notation we use to describe the superset block.

Based on the framework of Prob. 6.1, the control problem for superset block can be written as follows:

Problem 6.6 (Superset Junction Block). For $i = 1, \ldots, S$ and $j = 1, \ldots, M$,

$$
\begin{aligned}
\underset{\substack{x_i, u_i \\ x_j, u_j}}{\text{minimise}} \quad & \sum_{i=1}^{S} f_i(x_i, u_i) + \sum_{j=1}^{M} f_j(x_j, u_j) \\
\text{s.t.} \quad & A_{di} x_i + B_{di} u_i = \sum_{j \in \mathbf{M}_i^d} u_{ji,in}, \quad d \in \text{mat}_i^{out} \\
& A_j x_j + B_j u_j = v_j \\
& A_{i,loc} x_i + B_{i,loc} u_i = 0 \\
& A_{j,loc} x_j + B_{j,loc} u_j = 0 \\
& x_i \in X_i, u_i \in U_i \\
& x_j \in X_j, u_j \in U_j
\end{aligned}
\tag{6.15}
$$

where $i = 1, \ldots, S$ and $j = 1, \ldots, M$ respectively are the indices of supplier and customer elements. The first constraint represents the links between supplier and customer elements through material streams $d \in \text{mat}_i^{out}$, where A_{di}

Table 6.1. Notation for superset junction block

M	: Number of customer units in the second stage (units indexed by $j \in [1, \ldots, M]$)
S	: Number of supplier units in the first stage (units indexed by $i \in [1, \ldots, S]$)
mat_i^{out}	: Set of outgoing material streams of supplier unit i (streams indexed by $d \in \text{mat}_i^{out}$)
mat_j^{in}	: Set of incoming material streams of customer unit j (streams indexed by $q \in \text{mat}_j^{in}$)
$\mathbf{M}_i$	: Indices of customer units associated with supplier unit i
$\mathbf{S}_j$	: Indices of supplier units associated with customer unit j
$\mathbf{M}_i^d$	: Indices of customer units connected with material stream d in supplier unit i ($\mathbf{M}_i^d \subseteq \mathbf{M}_i$)
$\mathbf{S}_j^q$	: Indices of supplier units connected with material stream q in customer unit j ($\mathbf{S}_j^q \subseteq \mathbf{S}_j$)
K	: Current iteration index in the algorithm
$\mathbf{K}$	: Set of iterations $\{1, \ldots, K\}$
k	: Iteration index, $k \in \{1, \ldots, K\}$

and B_{di} represent the dth row of A_i and B_i. As discussed previously, the variable $u_{ji,in}$ represents the input demand from customer unit j to supplier unit i, while $\hat{v}_j$ is the demand that customer unit j receives from its customer units in the further downstream.$\square$

In what follows, we assume for simplicity that the local constraints $A_{i,loc}x_i + B_{i,loc}u_i = 0$ and $A_{j,loc}x_j + B_{j,loc}u_j = 0$ together with $x_i \in X_i, u_i \in U_i, i = \{1, \ldots, S\}$ and $x_j \in X_j, u_j \in U_j, j \in \{1, \ldots, M\}$ in Prob. 6.6 are sufficiently relaxed to absorb any demand imposed on respective unit element, *i.e.*, the sub- or master problems for supplier or customer units do not become infeasible for any v_i and v_j. The assumption, in turn, eliminates the need for generating *feasibility cuts* in the primal decomposition algorithm (see Section A.2, Appendix A). Consequently, to reduce the complexity of the description, we also omit these local constraints and implicitly assume that they always exist.

Algorithm 6.2 describes the generic procedure used for solving the superset block problem. The important step in the algorithm is the calculation of the update terms $\alpha_{ij,updt}(k)$ for individual junction blocks as listed in Table 6.2. For MIXER and MULTIFEED the algorithm operates exactly as per primal decomposition, with a single customer element, therefore the updates are set to 0. For SPLITTER block, the sum of demands $\hat{v}_{mS}$ for all customer elements $j \in [1, \ldots, M], m \neq j$ are treated as the parametric vector θ when referring to a customer element $m \in [1, \ldots, M]$. For MULTIPROD block, the vector of demands $\hat{v}_{1S}, \cdots, \hat{v}_{j-1S}, \hat{v}_{j+1S}, \cdots, \hat{v}_{MS}$ for all customer elements except m is treated as the parameter vector for element m. Thus, for both SPLITTER

Algorithm 6.2 (Superset Block Problem)

Step 0: Initialise *Set $K := 1$. Given $v_j = \hat{v}_j, j = 1, \ldots, M$, the demands for all second stage units. Assume an initial value of $u_{ji,in} \equiv u_{ji,in}^{(0)}$. Set $\hat{v}_{ji}^{(0)} = u_{ji,in}^{(0)}$ for all $i \in \boldsymbol{S}_j^q, q \in mat_j^{in}, j = 1, \ldots, M$.*

Step 1: Sub-problem $SP_i^{(K)}, i = 1, \ldots, S$: *At any iteration K, solve unit i's problem*

$$SP_i^{(K)} \quad \begin{cases} \underset{x_i, u_i}{minimise} \; \alpha_i \triangleq f_i(x_i, u_i) \\ \quad s.t. \quad A_{di}x_i + B_{di}u_i = \sum_{j \in M_i^d} \hat{v}_{ji}^{(K-1)}, \quad d \in mat_i^{out} \end{cases} \tag{6.16}$$

to obtain the optimal values of $x_i^{(K)}$, $u_{i,in}^{(K)}$, Lagrange multipliers $\lambda_i^{(K)}$ for linking equality constraints, and the objective function $\alpha_i^{(K)}$.

Set $z_{ij}^{(K)} \equiv \hat{v}_{ji}^{(K-1)}$, $j \in \boldsymbol{M}_i^d$, $d \in mat_i^{out}$, i.e., assume the demands from all master problems for all material streams $d \in mat_i^{out}$ is satisfied.

Step 2: Master Problem $SP_j^{(K)}, j = 1, \ldots, M$: *Use $\alpha_i^{(K)}, \lambda_{ij}^{(K)}$ and $z_{ij}^{(K)}$ to form a new optimality cut in unit j's problem $SP_j^{(K)}$. Solve the resulting problem*

$$SP_j^{(K)} \quad \begin{cases} \underset{x_j, u_j, \nu_j}{minimise} \; \alpha_j \triangleq \nu_j + f_j(x_j, u_j) \\ \quad s.t. \quad \nu_j \geq \sum_{i \in \boldsymbol{S}_j} \left\{ \alpha_{ij}^{(k)} + \alpha_{ij,updt}^{(k)} + \lambda_{ij}^{(k)}(z_{ij}^{(k)} - u_{ji,in}) \right\}, \quad k \in \boldsymbol{K} \\ \quad A_j x_j + B_j u_j = \hat{v}_j \end{cases}$$

$$\tag{6.17}$$

to obtain the solution $u_{ji,in}^{(K)}$ and $x_j^{(K)}$. The $\alpha_{ij,updt}^{(k)}$ are the approximate optimality cut updates used for updating the master problem j for perturbations in the demands from remaining other master problems $m \in [1, \ldots, M], m \neq j$. The updates are calculated by the first stage units $i = 1, \ldots, S$ and passed to the second-stage units $j = 1, \ldots, M$. Table 6.2 describes the specific formulation of these updates for all four junction blocks.

Step 3: Iterate/Terminate: *Terminate if the convergence criteria is satisfied, which is considered here as, for a given $\epsilon > 0$ and $i = 1, \ldots, S, j = 1, \ldots, M$,*

$$\left\| \left\{ x_i^{(K)}, u_i^{(K)}, x_j^{(K)}, u_j^{(K)} \right\} - \left\{ x_i^{(K-1)}, u_i^{(K-1)}, x_j^{(K-1)}, u_j^{(K-1)} \right\} \right\| \leq \epsilon \tag{6.18}$$

i.e., the solutions of sub-problems $SP_i^{(K)}$, $i = 1, \ldots, S$ and master problems $SP_j^{(K)}$, $j = 1, \ldots, M$ converge to a fixed point with tolerance ϵ. Else, set $\hat{v}_{ji}^{(K)} = u_{ji,in}^{(K)}$, $K := K + 1$, and return to Step 1. $\square$

---

and MULTIPROD blocks, a parametric update vector $\alpha_{Sj,updt}^{(k)}$ is calculated at each iteration for each of the customer elements $m = 1, \ldots, M$. In summary, Algorithm 6.2 provides a solution strategy for solving the two-stage control problems for each of the junction blocks in a distributed manner.

Table 6.2. Approximate optimality cut updates for junction blocks

---

MIXER $\alpha_{iM,updt}^{(k)} = 0$

MULTIFEED $\alpha_{iM,updt}^{(k)} = 0$

SPLITTER

$$\alpha_{Sj,updt}^{(k)} \triangleq \lambda_S^{(k)} \left(\sum_{\substack{m\in[1,...,M]\\m\neq j}} \hat{v}_{mS}^{(k-1)} - \sum_{\substack{m\in[1,...,M]\\m\neq j}} \hat{v}_{mS}^{(K-1)} \right), \quad k \in \mathbf{K}$$

(6.19)

where S is the single supplier unit, j is the customer unit for which the update is being calculated, and M is the last customer unit.

MULTIPROD

$$\alpha_{Sj,updt}^{(k)} \triangleq \left[\lambda_{S1}^{(k)}, \cdots , \lambda_{Sj-1}^{(k)}, \lambda_{Sj+1}^{(k)}, \cdots , \lambda_{SM}^{(k)} \right] \cdot$$
$$\left[\left(\hat{v}_{1S}^{(k-1)}, \cdots , \hat{v}_{j-1S}^{(k-1)}, \hat{v}_{j+1S}^{(k-1)}, \cdots , \hat{v}_{MS}^{(k-1)} \right)^T - \right.$$
$$\left. \left(\hat{v}_{1S}^{(K-1)}, \cdots , \hat{v}_{j-1S}^{(K-1)}, \hat{v}_{j+1S}^{(K-1)}, \cdots , \hat{v}_{MS}^{(K-1)} \right)^T \right],$$

$$k \in \mathbf{K} \quad (6.20)$$

where S is the single supplier unit, j is the customer unit for which the update is being calculated, and M is the last customer unit.

6.6 Solution of the Multi-Stage Problem

As the final step to solving Prob. 6.1, we need to ensure that all of the interconnected two-level junction block problems are solved iteratively in an appropriate sequence.

In what follows, we consider solving a sequence of three problems to develop the strategy in a constructive manner.

i. N-Units problem involving N series-connected units (Section 6.6.1)
ii. N-Units problem with an uncontrolled parameter in one or more of unit problems (Section 6.6.2)
iii. Main distributed control problem – Prob. 6.1 (Section 6.6.3)

6.6.1 Solution of N-Units Problem

We consider first an extension of the Two-Units problem to an N-Units problem as comprising N series-connected unit elements.

Problem 6.7 (N-Units Problem). For $i = 1, \ldots, N$

$$
\begin{aligned}
\operatorname*{minimise}_{\substack{u_i, x_i \\ i=1, \ldots, N}} \quad & \sum_{i=1}^{N} f_i(u_i, x_i) \\
\text{s.t.} \quad & A_i x_i + B_i u_i = v_i \quad i = 1, \ldots, N \\
& x_i \in X_i, u_i \in U_i \quad i = 1, \ldots, N.
\end{aligned}
\tag{6.21}
$$

where $v_i \equiv y_{i+1} = u_{i+1i,in}$ represents the demand to unit $i = 1, \ldots, N-1$. $\square$

A nested decomposition algorithm based on primal decomposition simply extends Algorithm 6.1 to repeatedly solve a sequence of $N-1$ two-units problems, one corresponding to each link between two successive unit elements. The sub-problem for each unit $i = 2, \ldots, N$ is considered as a master problem for a composite problem comprising all predecessor units 1 to $i-1$. The combined problem of units 1 to $i-1$ and unit i then becomes the sub-problem for unit $i+1$, and so on, until unit N is reached. The sub-problems of units $i = 1, \ldots, N$ are thus solved sequentially to construct a new optimality cut in the immediate next master problem. Once a complete iteration is finished, the procedure repeats starting from unit 1. In reference to Algorithm 6.1, the formulation of sub-problem SP_i at iteration K for $i = 2, \ldots, N$ along the sequence becomes as follows. We do not describe the complete algorithm for brevity.

Step i: Sub-problem SP_i, $i = 2, \ldots, N$: Use $\alpha_{i-1}^{(K)}, \lambda_{i-1}^{(K)}$ and $z_{i-1}^{(K)}$ to construct a new optimality cut in unit i's problem $SP_i^{(K)}$. Solve the resulting problem as:

$$
SP_i^{(K)} \quad
\begin{cases}
\operatorname*{minimise}_{x_i, u_i, \nu_i} \alpha_i \triangleq \nu_i + f_i(x_i, u_i) \\
\quad \text{s.t.} \quad \nu_i \geq \alpha_{i-1}^{(k)} + \lambda_{i-1}^{(k)}(z_{i-1}^{(k)} - u_{ii-1,in}), \quad k \in \mathbf{K} \\
\qquad\quad A_i x_i + B_i u_i = \hat{v}_i^{(K-1)}, \\
\qquad\quad x_i \in X_i, u_i \in U_i \\
\quad \text{where} \quad \mathbf{K} \triangleq \text{set of iterations}
\end{cases}
\tag{6.22}
$$

Set $z_i^{(K)} \equiv A_i x_i^{(K)} + B_i u_i^{(K)}$. Pass $\alpha_i^{(K)}, \lambda_i^{(K)}$ and $z_i^{(K)}$ to unit $i+1$.

6.6.2 Solution of N-Units Parametric Problem

We next consider an extension of the N-Units problem where the sub-problem of unit element 1 now contains an additional parameter vector $\theta \in \Theta$ while the remaining sub-problems $SP_i, i = 2, \ldots, N$ are same as in Prob. 6.7. Formally, the modified problem is:

Problem 6.8 (N-Units Problem with Parametric $SP_{1,\theta}$).

$$
\begin{aligned}
&\underset{x_i, u_i}{\text{minimise}} \sum_{i=1}^{N} f_i(x_i, u_i) \\
&\quad \text{s.t.} \quad A_i x_i + B_i u_i = v_i \quad i = 1, \ldots, N \\
&\qquad\qquad A_{1,\theta} x_1 + B_{1,\theta} u_1 = \theta \quad i = 1 \\
&\qquad\qquad x_i \in X_i, u_i \in U_i \quad i = 1, \ldots, N.
\end{aligned}
\tag{6.23}
$$

where $v_i \equiv y_{i+1} = C_{i+1} x_{i+1} + D_{i+1} u_{i+1}$ represents the interaction variable to unit i, $i = 1, \ldots, N - 1$.

The solution of the above problem is faced with the same challenge because the parametric Two-Units problem (Prob. 6.3) as the changes in θ in SP_1 results in a non-unique response to SP_2, and therefore, the non-unique response of all subsequent sub-problems to their immediate next master problem. Fortunately, an extension of the approximate cut update technique provides a method to resolve this. In this extension, we simply pass the update vector $\alpha_{1,updt}^{(\mathbf{K})}$ (of dimension K) from unit 1 to all units $i = 2, \ldots, N$ to modify the optimality cuts in their problems SP_i in a similar manner to sub-problem SP_2 in solving the parametric Two-Units problem. In the modified form of N-units algorithm, the sub-problem SP_i at iteration K is solved as follows.

Step i: Sub-problem SP_i, $i = 2, \ldots, N$: For each $k \in \mathbf{K}$, use $\alpha_{1,updt}^{(\mathbf{K})}$ to calculate $\alpha_{i-1,mod}^{(\mathbf{K})} \triangleq \alpha_{i-1}^{(\mathbf{K})} + \alpha_{1,updt}^{(\mathbf{K})}$, and solve the modified SP_i in Eq. 6.22, $i = 2, \ldots, N$ as:

$$
SP_i^{(K)} \left\{
\begin{aligned}
&\underset{u_i, x_i, \nu_i}{\text{minimise}} \; \alpha_i \triangleq \nu_i + f_i(u_i, x_i) \\
&\quad \text{s.t.} \quad \nu_i \geq \alpha_{i-1,mod}^{(k)} + \lambda_{i-1}^{(k)}(z_{i-1}^{(k)} - y_i), k \in \mathbf{K} \\
&\qquad\qquad A_i x_i + B_i u_i = \hat{v}_i^{(K-1)} \rightsquigarrow \lambda_i^{(K)} \\
&\qquad\qquad x_i \in X_i, u_i \in U_i \\
&\quad \text{where} \quad y_i \equiv u_{ii-1,in}
\end{aligned}
\right.
\tag{6.24}
$$

Note that we use the same update vector $\alpha_{1,updt}^{(\mathbf{K})}$ to all update all unit sub-problems SP_i, $i = 2, \ldots, N$. This communication of update vector can be made symmetric by assigning $\alpha_{2,updt}^{(\mathbf{K})} \triangleq \alpha_{1,updt}^{(\mathbf{K})}$ and repeating $\alpha_{i,updt}^{(\mathbf{K})} \triangleq \alpha_{i-1,updt}^{(\mathbf{K})}$ for all $i = 3, \ldots, N$, and using $\alpha_{i-1,mod}^{(\mathbf{K})} \triangleq \alpha_{i-1}^{(\mathbf{K})} + \alpha_{i-1,updt}^{(\mathbf{K})}$.

We consider next a further extension where instead of just sub-problem SP_1, one or more of other sub-problems are also parameterized with parameter vectors $\theta_i \in \Theta_i$. The approximate cut update technique can as well be extended to solve this problem in an analogous manner.

For SP_{1,θ_1} we still continue to propagate the approximate cut update $\alpha_{1,updt}^{(K)}$ to all units $i = 2, \ldots, N$. In addition, for any sub-problem SP_{i,θ_i}, $i = 2, \ldots, N - 1$, we also generate a separate approximate cut update that reflects the effects of parametric variations in its parameter vector θ_i. This can be written as: $\alpha_{i,loc,updt}^{(k)} \triangleq -\lambda_{i,\theta_i}^{(k)^T} (\theta_i^{(K)} - \theta_i^{(k)})$, $k \in \mathbf{K}$ where λ_{i,θ_i} is the Lagrange multiplier associated with the constraint $A_{i,\theta_i} x_i + B_{i,\theta_i} u_i = \theta_i$.

We then simply add this update to the update received from previous sub-problem $SP_{i-1,\theta_{i-1}}$ and pass the composite update to the next sub-problem $SP_{i+1,\theta_{i+1}}$, i.e., $\alpha_{i,updt}^{(K)} \triangleq \alpha_{i-1,updt}^{(K)} + \alpha_{i,loc,updt}^{(K)}$. With this modification the sub-problem $SP_{i,\theta_i}^{(K)}$, $i = 2, \ldots, N$ at iteration K is now solved as follows.

Calculate $\alpha_{i-1,mod}^{(K)} \triangleq \alpha_{i-1}^{(K)} + \alpha_{i-1,updt}^{(K)}$. Use $\alpha_{i-1,mod}^{(K)}$ in solving

$$
SP_i^{(K)} \begin{cases}
\underset{u_i,x_i,\nu_i}{\text{minimise}} \ \alpha_i \triangleq \nu_i + f_i(u_i, x_i) \\
\quad \text{s.t.} \quad \nu_i \geq \alpha_{i-1,mod}^{(k)} + \lambda_{i-1}^{(k)}(z_{i-1}^{(k)} - y_i), k \in \mathbf{K} \\
\qquad \quad A_i x_i + B_i u_i = \hat{v}_i^{(K-1)}, \rightsquigarrow \lambda_i^{(K)} \\
\qquad \quad A_{i,\theta_i} x_i + B_{i,\theta_i} u_i = \theta_i^{(K)} \rightsquigarrow \lambda_{i,\theta_i}^{(K)} \\
\qquad \quad x_i \in X_i, u_i \in U_i \\
\quad \text{where} \quad y_i \equiv u_{ii-1,in}
\end{cases} \tag{6.25}
$$

Using $\lambda_{i,\theta_i}^{(K)}$, calculate the approximate cut update to be passed to the next unit $i + 1$ as $\alpha_{i,updt}^{(K)} \triangleq \alpha_{i-1,updt}^{(K)} + \alpha_{i,loc,updt}^{(K)}$ where

$$
\alpha_{i,loc,updt}^{(k)} \triangleq \lambda_{i,\theta_i}^{(K)^T} (\theta_i^{(K)} - \theta_i^{(k)}), k \in \mathbf{K}
$$

6.6.3 Solution of Distributed Control Problem (Prob. 6.1)

We can now develop the distributed coordination algorithm for solving main distributed problem (Prob. 6.1) for an arbitrary, acyclic process network. To define the interactions between unit elements more systematically, we first develop an indexing of unit elements in the P-Graph.

Assume the process contains S different stages with each stage containing possibly one or more unit elements. The word *stage* refers to a typical processing task, such as the primary reaction, separation, *etc*. Each such stage may contain a number of unit elements of similar processing capabilities. We then use the following procedure to assign an index (n, s) to all unit elements.

We first assign the stage index 1 to all unit elements that use the main raw-materials as their feedstocks, *i.e.*, the input-set mat^{in} for these unit elements comprise only the main raw-materials. All unit elements connected to these first stage unit elements are then assigned the stage index 2. The assignment is thus repeated until the unit elements in the terminal stage reached whose output-set mat^{out} comprise only the end-products. These elements receive the stage index S. In this process of assigning stage indices, if an element receives two or more different indices, because of it being connected to unit elements from two or more stages, then the highest stage index among all is used. Next, within each stage, the units are numbered from 1 to a maximum value (called N_s) that depends on the unit elements contained in that stage. We use a simple rule of progressing from top to bottom in the P-Graph to define this unit number. At the end of this indexing, each unit element thus receives an index (n, s) where s refers to the stage index and n as the number of the unit element within that stage.

We next use the notation described in Table 6.3 to define the solution procedure for Prob. 6.1. Note that for any stage s, S_s^- and S_s^+ denote the stages preceding and succeeding to stage s and comprise unit elements linked to any unit element in stage s. Note also that as per above indexing rules at least one unit in any stage s must be linked to stage $s-1$ as well as stage $s+1$. Unit elements in stage s may also be linked to other stages in S_s^- and S_s^+. The set $\mathbf{M}_{(n,s)}$ hence encompasses the indices of unit elements in stages S_s^+ which are connected to unit element (n, s) in stage s. The set $\mathbf{M}_{(n,s)}^d$ then denotes the subset of unit elements within $\mathbf{M}_{(n,s)}$ that are connected to element (n, s) through the material stream $d \in \mathrm{mat}_{(n,s)}^{out}$. Similar interpretation can be given for $\mathbf{S}_{(n,s)}$ and $\mathbf{S}_{(n,s)}^q$.

The nested solution procedure for Prob. 6.1 extends the algorithm for superset block (Algorithm 6.2) by using the results from parametric N-unit problems from the previous subsection. Algorithm 6.3 describes this distributed algorithm. As expected, the important step in the algorithm is to compute the approximate cut update terms $\alpha_{(n,s)j,updt}^{(K)}$ to be passed between stages while taking into account the specific type of junction block by which the associated unit element is connected to other unit elements.

Table 6.3. Notation for distributed coordination algorithm

S	: Number of process stages (stages indexed by $s \in [1, \ldots, S]$)
$\mathbf{S}_s^-$	: Indices of stages preceding to stage s
$\mathbf{S}_s^+$	: Indices of stages succeeding to stage s
N_s	: Number of units in stage n (units indexed by $n \in [1, \ldots, N_s]$)
(n, s)	: Index of unit n in stage s, $n \in [1, \ldots, N_s]$, $s \in [1, \ldots, S]$
$\mathrm{mat}_{(n,s)}^{out}$	: Set of outgoing material streams of unit (n, s) (indexed by $d \in \mathrm{mat}_{(n,s)}^{out}$)
$\mathrm{mat}_{(n,s)}^{in}$	: Set of incoming material streams of unit (n, s) (indexed by $q \in \mathrm{mat}_{(n,s)}^{in}$)
$\mathbf{M}_{(n,s)}$	: Indices of units in $\mathbf{S}_s^+$ connected with unit (n, s)
$\mathbf{M}_{(n,s)}^d$	: Indices of units in $\mathbf{S}_s^+$ connected with outgoing stream d of unit (n, s) $(\mathbf{M}_{(n,s)}^d \subseteq \mathbf{M}_{(n,s)})$
$\mathbf{S}_{(n,s)}$	: Indices of units in $\mathbf{S}_s^-$ connected with unit (n, s)
$\mathbf{S}_{(n,s)}^q$	: Indices of units in $\mathbf{S}_s^-$ connected with incoming stream q of unit (n, s) $(\mathbf{S}_{(n,s)}^q \subseteq \mathbf{S}_{(n,s)})$

Algorithm 6.3 (Distributed Coordination Algorithm)

Step 0: Set $K := 1$. Given $v_{(n,S)} = \hat{v}_{(n,S)}$, the demands for unit elements (n, S), $n = 1, \ldots, N_S$ in the terminal stage S. For all $s = [1, \ldots, S]$ and $n = [1, \ldots, N_s]$, assume an initial value of $u_{(n,s)i,in} \equiv u_{(n,s)i,in}^{(0)}$, and for all $i \in S_{(n,s)}$ set $\hat{v}_{(n,s)i}^{(0)} = u_{(n,s)i,in}^{(0)}$.

Step s, $s = 1, \ldots, S$: Sub-problem $SP_{(n,s)}^{(K)}, n = 1, \ldots, N_s$: Use $\alpha_{i(n,s)}^{(K)}$, $z_{i(n,s)}^{(K)}$, $i \in S_{(n,s)}$ to form a new optimality cut in unit (n, s)'s problem $SP_{(n,s)}^{(K)}$. Solve the resulting problem

$$
SP_{(n,s)}^{(K)} \left\{
\begin{aligned}
&\underset{\substack{x_{(n,s)}, u_{(n,s)}, \\ \nu_{(n,s)}}}{minimise} \ \alpha_{(n,s)} \triangleq \nu_{(n,s)} + f_{(n,s)}(x_{(n,s)}, u_{(n,s)}) \\
&s.t. \quad \nu_{(n,s)} \geq \sum_{i \in S_{(n,s)}} \Big\{ \alpha_{i(n,s)}^{(k)} + \alpha_{i(n,s),updt}^{(k)} + \\
&\qquad\qquad\qquad \lambda_{i(n,s)}^{(k)} (z_{i(n,s)}^{(k)} - u_{(n,s)i,in}) \Big\}, \quad k \in K \\
&\quad A_{d(n,s)} x_{(n,s)} + B_{d(n,s)} u_{(n,s)} = \sum_{j \in M_{(n,s)}^d} \hat{v}_{j(n,s)}^{(K-1)}, \quad d \in mat_{(n,s)}^{out}
\end{aligned}
\right.
$$

$$(6.26)$$

to obtain the optimal values of $x_{(n,s)}^{(K)}, u_{(n,s)i,in}^{(K)}$, optimal Lagrange multipliers $\lambda_{(n,s)}^{(K)}$ associated with the linking equality constraints and the optimal objective value $\alpha_{(n,s)}^{(k)}$. Note that $A_{d(n,s)}$ and $B_{d(n,s)}$ denote the dth row in constraint matrices $A_{(n,s)}$ and $B_{(n,s)}$ associated with unit (n, s). The solution $u_{(n,s)i,in}^{(K)}$ defines the demands sent by unit (n, s) to all linked supplier elements $S_{(n,s)}$ at the next iteration.

Take $z_{(n,s)j}^{(K)} \equiv \hat{v}_{j(n,s)}^{(K-1)}$, $j \in M_{(n,s)}^d$, $d \in mat_{(n,s)}^{out}$. Set $\lambda_{(n,s)j}^{(K)} \equiv \lambda_{(n,s)}^d$, $\forall j \in M_{(n,s)}^d$ where $\lambda_{(n,s)}^d$ denotes the dth element in $\lambda_{(n,s)}^{(K)}$. Calculate the aggregate cut update $\alpha_{(n,s)j,updt}$ as,

$$
\alpha_{(n,s)j,updt}^{(K)} \triangleq \sum_{i \in S_{(n,s)}^q} \Big\{ \alpha_{i(n,s),updt}^{(K)} \Big\} + \alpha_{(n,s)j,loc,updt}^{(K)} \tag{6.27}
$$

where $\alpha_{(n,s)j,loc,updt}^{(K)}$ denotes the approximate cut updates that unit element (n, s) generates locally for parametric demand variations from customer elements $j \in M_{(n,s)}$ while $\alpha_{i(n,s),updt}^{(K)}$ are the approximate cut updates that it receives from supplier elements $i \in S_{(n,s)}$.

Step N+1: Iterate/Terminate: Terminate if the specific convergence criteria is satisfied, which is considered as, for a given $\epsilon > 0$ and for all $s = [1, \ldots, S]$ and $n = [1, \ldots, N_s]$,

$$
\left\| \Big\{ x_{(n,s)}^{(K)}, u_{(n,s)}^{(K)} \Big\} - \Big\{ x_{(n,s)}^{(K-1)}, u_{(n,s)}^{(K-1)} \Big\} \right\| \leq \epsilon \tag{6.28}
$$

Else, set $\hat{v}_{(n,s)i}^{(K)} = u_{(n,s)i,in}^{(K)}$, $K := K + 1$ and return to Step 1. $\square$

Note the bidirectional nature of information flow between unit elements in Algorithm 6.3 – the flowrate demands for feedstocks (in the form of $u^{(K)}_{(n,s)i,in}$) flow backwards and the resulting supply proposals (in the form of optimality cuts) flow forwards.

6.7 Implementation and Numerical Examples

The distributed coordination algorithm (Algorithm 6.3) developed in the previous section was implemented using MATLAB® software, the details of the implementation are provided in Appendix C. In what follows, we simply describe a few numerical examples to illustrate the different features of the algorithms discussed in the previous sections. An application of Algorithm 6.3 to an industrial-scale, multipurpose process problem is discussed in the next chapter.

Example 6.9. The first example illustrates the approximate optimality cut update technique described in Section 6.5.2. Consider the following problem.

$$\begin{array}{cl} \underset{x_{1,1},x_{1,2},x_{1,3},x_2}{\text{minimise}} & x^2_{1,1} + x^2_{1,2} + x^2_{1,3} + x^2_2 + x_{1,1} + x_{1,2} + x_{1,3} \\ \text{s.t.} & \begin{array}{l} x_{1,1} + x_{1,2} + x_{1,3} + x_2 = 3 \\ 2.5x_{1,1} + 0.7x_{1,2} + 1.6x_{1,3} + \theta = 5 \end{array} \end{array} \qquad (6.29)$$

where $x_{1,\cdot}$ and x_2 refer to the local variables of unit elements 1 and 2, while θ is the parameter vector local to unit element 1. The formulation of the associated sub-problems at iteration K in the form of Algorithm 6.1 can be written as follows:

$$SP^{(K)}_{1,\theta} \begin{cases} \underset{x_{1,1},x_{1,2},x_{1,3}}{\text{minimise}} \ \alpha_1 \triangleq x^2_{1,1} + x^2_{1,2} + x^2_{1,3} + x_{1,1} + x_{1,2} + x_{1,3} \\ \text{s.t.} \quad \begin{array}{l} x_{1,1} + x_{1,2} + x_{1,3} = 3 - \hat{x}^{(K-1)}_2 \rightsquigarrow \lambda^{(K)}_1 \\ 2.5x_{1,1} + 0.7x_{1,2} + 1.6x_{1,3} + \theta = 5 \end{array} \end{cases} \qquad (6.30)$$

and

$$SP^{(K)}_2 \begin{cases} \underset{\nu_2,x_2}{\text{minimise}} \ \alpha_2 \triangleq \nu_2 + x^2_2 \\ \text{s.t.} \quad \nu_2 \geq \alpha^{(k)}_1 + \lambda^{(k)}_1 (x^{(k)}_{1,1} + x^{(k)}_{1,2} + x_2 - 3), k \in \mathbf{K} \end{cases} \qquad (6.31)$$

where λ_1 is the Lagrange multiplier. Assume $\theta = 1$ when the algorithm is started and that it changes to $\theta = 5$ at iteration 5 and remains the same thereafter. If the approximate cut updates are not used when θ changes, then the master problem will retain the optimality cuts from previous iterations and result in an incorrect solution. Fig. 6.11 illustrates this effect by comparing the value function α_2 as a function of x_2 for three different scenarios: Case

A – when no updates are used; Case B – when the algorithm is restarted at iteration 6 after θ is changed; and, Case C – when the approximate cut updates are applied. Note that the parametric effect results in an incorrect solution in case A, while the use of approximate cut updates in case C gives the same result as case B where the algorithm is restarted after the change in θ.

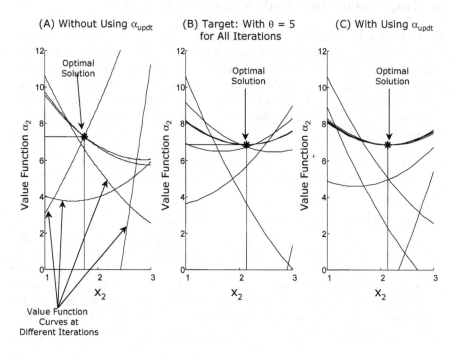

Fig. 6.11. Example 6.9: Effect of using approximate cut updates α_{updt} – (A): Without α_{updt} results in an incorrect solution, (B): Target solution – restarting the algorithm with $\theta = 5$, (C): With α_{updt} – results in a correct Solution

The next four examples refer to application of the Superset block algorithm (Algorithm 6.2) to four different process networks shown in Fig. 6.12. The data for matrices Q_i, A_i, $A_{i,loc}$ and vectors c_i, B_i, $B_{i,loc}$ for these examples are given in Appendix C.

Example 6.10. (MIXER) The first example refers to a MIXER block with three unit elements. Table C.2 in Appendix C shows the progress of iterations during an execution of Algorithm 6.2. The variables $u_{(n,s,z)}$ and $x_{(n,s,z)}$ therein refer to the zth element of the input and state vectors of sub-problem $SP_{(n,s)}$, while the last column shows the solution obtained for overall problem using a centralised algorithm. As can be verified, the solution obtained via distributed

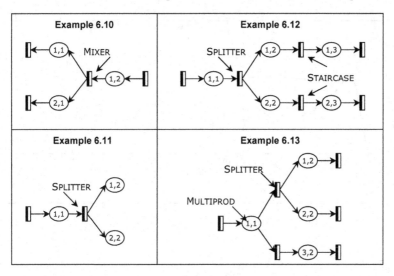

Fig. 6.12. Process configuration for Examples 6.10 to 6.13

algorithm matches with that from centralised algorithm within a predefined tolerance.

Example 6.11. (SPLITTER) The next problem refers to a SPLITTER block with three unit elements as shown in Fig. 6.12. Note that for unit $(1,1)$, the demand from unit $(2,2)$ becomes the uncontrollable parameter θ when referred to $(1,2)$. Similarly, the demand from unit $(1,2)$ becomes the parameter θ when referred to unit $(2,2)$. The approximate cut update technique resolves these parametric effects by updating the optimality cuts in the sub-problems of units $(1,2)$ and $(2,2)$. Table C.4 in Appendix C shows the progress of iterations for Algorithm 6.2, where again the last column confirms that the resulting solution matches with that from centralised algorithm within a predefined tolerance.

To demonstrate an additional feature of approximate cut updates, we change the terminal demands from 10 to 20 deviation units for units $(1,2)$ and $(2,2)$ while the execution of the algorithm is in progress. Fig. 6.13 shows that the algorithm is able to pick up the change and converge to a new optimum.

Example 6.12. (SPLITTER-STAIRCASE) The next example illustrates the nesting of junction blocks in a SPLITTER -STAIRCASE process, where the STAIRCASE block refers to a Two-Units process as a special case of the MIXER or MULTIFEED blocks.

As per the information flow described in Section 6.3, the demand $\hat{v}_{(1,3)(1,2)}$ from unit $(1,3)$ to $(1,2)$ propagates backwards to demand $\hat{v}_{(1,2)(1,1)}$ that unit $(1,2)$ sends to unit $(1,1)$. The same applies to units $(2,3)$ and $(2,2)$. The demands from terminal units $(1,3)$ and $(2,3)$ thus parameterize the sub-problem

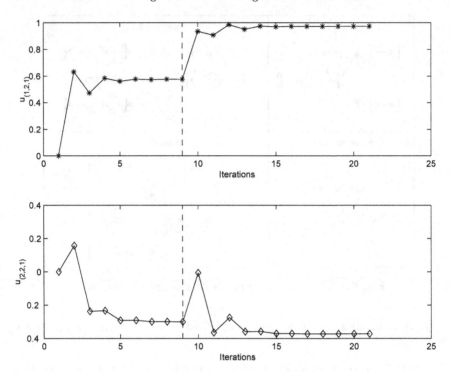

Fig. 6.13. Example 6.11: Effects of change in terminal demands for units $(1, 2)$ and $(2, 2)$

of unit $(1, 1)$. The use of propagation of approximate cut updates along the network resolves these parametric effects by updating the optimality cuts in all four units $(1, 2)$, $(1, 3)$, $(2, 2)$ and $(2, 3)$ as described in Algorithm 6.3. Table C.6 in Appendix C summarises the progress of iterations and a comparison with the equivalent centralised solution.

Fig. 6.14, similar to Fig. 6.13, shows the ability of the algorithm to pick up a change in demands for units $(1, 3)$ and $(2, 3)$ from 10 to 20 deviation units. The plots refer to the input demands $u_{(1,2,1)}, u_{(2,2,1)}, u_{(1,3,1)}, u_{(2,3,1)}$ that the second and third stage units $(1, 2), (2, 2), (1, 3)$ and $(2, 3)$ request from their supplier units. As can be seen, the algorithm converges to a new optimum.

Example 6.13. (MULTIPROD-SPLITTER) The final example illustrates the nesting of a MULTIPROD block with a SPLITTER block. The example shows the combined parametric effects from MULTIPROD and SPLITTER blocks within a single problem. The demand from unit $(3, 2)$ becomes the parameter θ for unit $(1, 1)$ when referring to a combined problem of units $(1, 2)$ and $(2, 2)$. Units $(1, 2)$ and $(2, 2)$ thus both receive the same approximate cut update of the MULTIPROD type (Eq. 6.20) for demand variations from unit $(1, 3)$. Unit $(3, 2)$ similarly also receives a cut update of MULTIPROD type (Eq. 6.20)

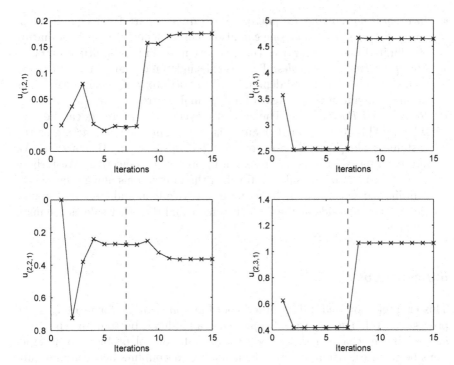

Fig. 6.14. Example 6.12: Effects of change in terminal demands for units $(1,3)$ and $(2,3)$

for a combined demand from units $(1,2)$ and $(2,2)$. In addition, the demands from units $(1,2)$ and $(2,2)$ also parameterize unit $(1,1)$'s sub-problem for each other's demand. Hence, these two units also receive an additional SPLITTER type cut update of the form Eq. 6.19.

Tables C.8 in Appendix C summarises the progress of iterations and the convergence to the optimal solution obtained by a centralised algorithm.

6.8 Future Extensions

In the course of developing the distributed algorithm, we made various assumptions that helped us simplify the discussions. The algorithm can be extended and generalised to a wider class of problems if one or more of these assumptions are relaxed. For example:

- *Infeasible Sub-problems:* The assumption that constraints $xi \in X_i$, $u_i \in U_i$ or $A_{i,loc}x_i + B_{i,loc}u_i = 0$ are sufficiently relaxed to allow unit i to accept any product demand $\hat{v}_i$ from downstream units can be relaxed by using so-called *feasibility restoration technique* for primal decomposition concept (Grothey, Leyffer & McKinnon 1999).

- *Incorporating Linking Inequality Constraints:* Apart from equality constraints $A_i x_i + B_i u_i = \hat{v}_i$, one can also include inequalities such as sharing of a limited quota of services (*e.g.*, energy flow) linking multiple units.
- *Multiple Demand Variables:* The case of singleton demand in $\hat{v}_i$ can be extended to a vector-valued demand, *e.g.*, a trajectory of demand variations in time domain in the context of an optimal control problem.
- *Recycle and By-products:* The case of recycle or by-products can be considered. This requires further analysis as the unit elements acting as the customers along recycle are now situated upstream in the process. An approach based on classical research in process flowsheeting (Westerberg *et al.* 1979) can be considered in which the interactions along recycles are coordinated separately by breaking the recycle loop and treating the unit elements on one side as the final customers and the other side as the main suppliers.

6.9 Summary

This chapter proposed a distributed coordination strategy for reconfigurable process control. The key to the approach is a modular, bottom-up type problem solving mechanism that solves the overall control problem by interactions between (distributed) unit elements. The decoupling between unit subproblems in the solution technique enables the introduction of new unit elements. The unit elements are also able to respond to local disturbances dynamically and adjust their settings (as shown via demand changes in Examples 6.11 to 6.13). We also note that – in line with typical supply chain behaviour – each unit element, acting as the customer of its feedstock demands, attempts to coordinate these demands among potential supplier elements. A propagation of this demand distributed across the process scheme ensures the process responds to changes in the demands in a dynamic and incremental manner.

Part III

An Assessment of the DRPC Approach

7

Application of Distributed Coordination Approach – A Case Example

7.1 Introduction

To illustrate the application of distributed coordination approach, we now describe an example of an industrial-scale, multipurpose process plant. The example is derived from a similar example in Friedler *et al.* (1992) and reflects largely the characteristics of modern process plants in petrochemicals, polymers and chemicals industries, except that, for simplicity, we omit the complexities of recycles or byproducts. These limitations however do not impede the generality of discussions in this chapter. The multipurpose nature of the example allows us to analyse a number of potential production scenarios that can be expected to arise in this class of industry in future. In this sense the example also reflects the long-term vision of a highly reconfigurable process control system and shows that it can be developed using a distributed approach. The system has developed in sufficient detail that it might be used as a benchmark problem.

This chapter is structured as follows. The next section describes the example process considered in this chapter. Section 7.3 then introduces the problem formulation in terms of six different production scenarios used for analysis. The subsequent three sections then apply the developments from previous chapters to example process and these scenarios to illustrate how the proposed distributed coordination would operate under these conditions.

7.2 Process Description

The multipurpose process considered as example comprises 18 process units, each capable of performing one or more processing tasks. Fig. 7.1 diagrammatically shows the initial physical layout of the process, which might, for example, be used for polymer, polyester and some petrochemical products. The process is able to produce three products A, B and C. Fig. 7.2 shows

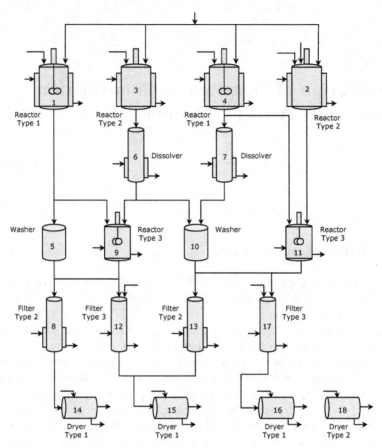

Fig. 7.1. Process layout for multipurpose process example

the initial product recipes for these products in terms of the sequence of processing tasks required to convert raw-materials to end-products. The oblong symbols therein represent the processing tasks while the rectangles represent the materials. The flow of materials is thus from top to bottom. Table 7.1 lists the structure of processing tasks in terms of the associated unit operation and the input and output materials for each task. Note that the recipes for all three products are of *non-linear* type, *i.e.*, there exists more than one task sequence that can produce the same end-product. The dark-lined sequence in each case is the preferred sequence over others.

We note that the tasks in Table 7.1 are not assigned to any processing units at this stage yet. Later in Section 7.4 we consider three different combinations of these 'initial' physical layout and product recipes to understand how the process of managing task assignment, *i.e.*, recipe mapping, works and the various physical and product issues that surround it.

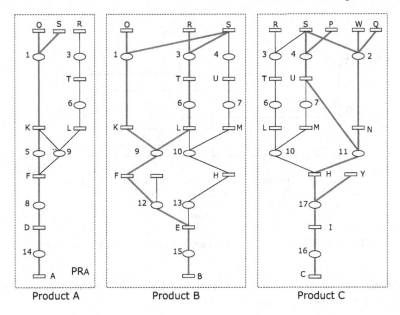

Fig. 7.2. Product recipes for products A, B, and C (the dark lines show the preferred task sequence

As can be seen we have omitted services in both Figs 7.1 and 7.2 in order to simplify the discussions, and to also focus on the key aspect of demand-pull

Table 7.1. List of processing tasks in multipurpose process example

No.	Type	Inputs	Outputs
1.	Reactor	O,S	K
2.	Reactor	S,W,Q	N
3.	Reactor	S,R	T
4.	Reactor	S,P	U
5.	Washer	K	F
6.	Dissolver	T	L
7.	Dissolver	U	M
8.	Filter	F	D
9.	Reactor	K,L	F
10.	Washer	L,M	H
11.	Reactor	U,N	H
12.	Filter	F,G	E
13.	Filter	H	E
14.	Dryer	D	A
15.	Dryer	E	B
16.	Dryer	I	C
17.	Filter	H Y	I
18.	Dryer	X	B

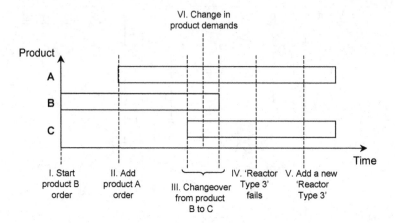

Fig. 7.3. Schedule of product campaigns and example scenarios

type behaviour of process elements. The description in this chapter can simply be extended to also cover service flows.

7.3 Problem Description

Our main purpose for the example considered in this chapter is to analyse the ability of a process control system to respond to a change in plant condition that demands a level of reconfigurability from process operations. To perform this analysis systematically, we propose a set of production scenarios representing the type of changes or disturbances that can be expected to arise in future process operations. These scenarios, while not fully exhaustive, illustrate a number of the features of the proposed distributed approach in a best possible way. Fig. 7.3 shows a time-line for these scenarios within a production run. They are defined as follows.

- **Scenario I – Start product** B **order:** Assume the process is idle at start, and that a new order for product B arrives. A campaign for producing B is initiated (This scenario should illustrate the process of integrating product recipe information in developing a process scheme);
- **Scenario II – Add product** A **order:** While B is being produced, assume an order for product A also arrives and is initiated immediately (This scenario should illustrate the process of managing change, in particular for those units which can be involved in both process schemes).
- **Scenario III – Changeover from product** B **to** C: While the order for B is nearing to its completion, assume an order for C arrives and is initiated in parallel to A and B. On completion, the order for B is stopped and removed (This scenario, similar to scenario II, should depict

Table 7.2. Links between scenarios and reconfigurability requirements

		Scenario					
		I	II	III	IV	V	VI
Requirements	Product/process diversity	✓	✓	✓			
	Easy modifiability		✓	✓	✓		
	Responsiveness		✓	✓	✓		✓
	Fault-tolerance					✓	

the process of changeover but in an opposite way, *i.e.*, the product B is now also removed).

- **Scenario IV – 'Reactor Type 3' fails:** Assume at this stage one of the 'reactor type 3' fails, and is unable to perform its task. (This scenario should illustrate the ability of the system to tackle failures and support graceful degradation of performance);
- **Scenario V – Add a new 'Reactor Type 3':** Finally, assume a new 'reactor type 3' is added in its place or assume the failed unit recovers after repair (This scenario should illustrate the modifiability of the network to support inclusion of a new facility, *e.g.*, a new process unit in this case).

The above scenarios illustrate the qualitative aspects of reconfiguration, which in the distributed case refer to architectural and interaction issues. In addition, a sixth scenario is considered below to demonstrate the quantitative aspect of the system to respond to a control change or disturbance.

- **Scenario VI – Change in product demands:** Assume all three products are being produced at the same time (*e.g.*, during changeover from B to C) and that the demands for all three change by 10 deviation units from their current demand set-points (This scenario should illustrate the responsiveness of the system in terms of propagation of demand changes to whole process network).

The above scenarios directly relate to four RPC requirements provided in Chapter 2 as shown in Table 7.2.

7.4 Application of the DRPC Approach

We now describe the distributed approach applied to this example. The description below follows the outline of developments in the previous three chapters, *i.e.*, (i) identification, (ii) organisation from Chapter 4 and (iii) interaction behaviour of process elements from Chapter 5 and (iv) coordination of their distributed process parameters from Chapter 6.

7.4.1 Identification of Process Elements

The identification of process elements is carried out based on the physical structure of the process.

- *Unit Elements:* Each process unit in Fig. 7.1 is associated with a separate unit element in the control architecture. In total, this results in 18 unit elements, each having a capability to perform one or more processing tasks from Table 7.1. The exact number of task(s) that a unit element performs is varied between a single task or multiple tasks as discussed in the next subsection.
- *Header Elements:* Each piping network connecting unit elements in subsequent stages in Fig. 7.1 is represented by a separate header element. This basically results in a header element for each raw-material, intermediate material and end-product. However, in a more flexible layout, as shown later in Figs. 7.5 and 7.6, the header elements can also be associated with more than one material types. We note that not all piping segments in the process need to be identified as header elements (*e.g.*, the connection between units 3 and 6 in Fig. 7.1) if their role is purely to connect two or more unit elements with no added decisions about process or routing flexibility.
- *Service Elements:* While services are omitted from discussions here, the suppliers of each service used by unit or header elements in Fig. 7.1 can be represented by an appropriate service element.
- *Product Elements:* Each customer order for any of the three products is represented by a product element. All three product elements can thus coexist in the process as in scenario III.

The process elements identified above are defined with their data models and control functions as shown in Fig. 4.3. We however omit these details and limit our focus onto their organisation and interaction behaviour.

7.4.2 Organisation of Process Elements

Similar to identification, the organisation of process elements mirrors their physical involvement in the process. In particular, each unit element is defined by the header elements that it is connected with, and each header elements by the unit or service elements that it connects together. In order to understand how this physical structure and interconnection of elements supports reconfigurability, we consider below how changing the flexibility available in the local design of unit, header or product elements can affect this property.

- *Unit Elements:* The capability of unit elements is varied between each being able to perform: (i) a single task, or (ii) multiple tasks, where a *task*, as defined in Section 5.2, refers to a unit operation (*e.g.*, reaction) with its associated materials and services.

Table 7.3. Variations in the organisation of process elements

	Recipe mapping Approach	Unit Element Capability	Physical Layout
Case 1	Product-centric	Single task	Fixed
Case 2	Unit-centric	Single task	Full
Case 3	Unit-centric	Multiple tasks	Flexible

- *Header Elements:* The capability of header elements is varied by changing their flexibility to interconnect process units between: (i) *fixed* connectivity, where connectivity is limited as in Fig. 7.1, (ii) *full* connectivity, where all unit elements can be connected to all other unit elements, and (iii) *flexible* connectivity, on the spectrum between fixed and full connectivity, where connectivity is enhanced by increasing the number of possible connections between unit elements in the fixed layout.
- *Product Elements:* The capability of product elements is varied by changing their their involvement in the process based on the approach used for recipe mapping, (see Section 5.2), *i.e.*, : (i) *product-centric approach*, where product elements are supplied with (non-linear) product recipes shown in Fig. 7.2, and (ii) *unit-centric approach*, where product elements are not supplied with any recipe at all, but this information is defined as part of the design of unit elements themselves. In the former, the product elements centrally assign processing tasks to unit elements, while in the latter the unit elements themselves select the tasks based on recipe information supplied to them.

The above variations thus entail different options by which the elements can be organised within the overall system. We consider, in particular, three such combinations characterised in Table 7.3 and represented in Fig. 7.4 (case 1), Fig. 7.5 (case 2) and Fig. 7.6 (case 3). The oblong symbols therein represent the unit elements, the rectangles represent the materials, and the lines connecting oblong and rectangle symbols are possible connections between unit elements. From these figures, we can make the following observations.

- In all three cases, it is assumed that the unit elements are not defined with any *a priori* information about other unit elements they may be connected with. They acquire this information from the associated header elements during synthesis phase.
- As compared to fixed layout in Fig. 7.4, the unit elements in the full (Figs. 7.5) or flexible (Fig. 7.6) layouts are defined with the exact set of materials they need to consume or produce. Embedding this additional information however does not fix the process schemes for either case as there exists multiple combinations of unit elements (in Fig. 7.5) or their

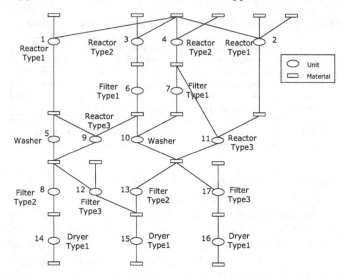

Fig. 7.4. Process layout: Case 1 - single task, fixed connectivity

selection of local tasks (in Fig. 7.6) which can produce the same end-product. The selection of a specific process scheme from these combinations occurs via distributed interactions between unit elements together with the associated product, header and service elements.

- In the full connectivity layout (Fig. 7.5), there exists no clear distinction between different header elements, rather the whole network can be seen as a single header element comprising multiple sub-networks connecting individual process stages. Note that a layout of this nature would be rare to find in reality however it shows the possibility of interconnections in so-called *pipeless* plants where the header elements can be thought as the material carrying equipment being moved around the plant.
- The flexible layout in Fig. 7.6 assumes that the unit elements are capable of performing multiple tasks. This feature in turn leads to a reduction in the unit element types from 18 in Figs. 7.4 or 7.5 to 8 in Fig. 7.6. As discussed later, this multipurpose capability combined with the flexible process layout in Fig. 7.6 helps enhance the reconfigurability of the control system to deal with changes or disturbances in a distributed way.

7.4.3 Interaction Behaviour of Process Elements

The process elements interact based on the interaction model presented in Chapter 5. With reference to the six scenarios described earlier, the *identify* phase leads to interpreting the effects of change or disturbance into specific requirements for reconfiguration. Where required (as in scenarios I, II, III), a new product element is also created to impose these requirements onto other

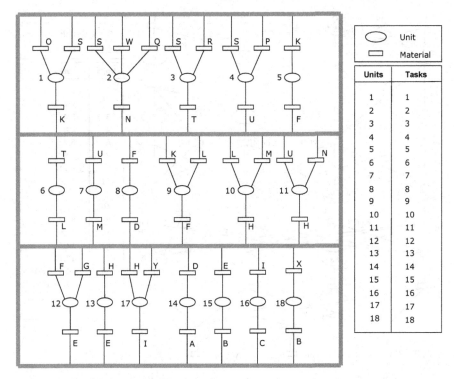

Fig. 7.5. Process layout: Case 2 - single task, full connectivity

process elements. The *define* and *reconfigure* phases then involve the unit elements in the process together with associated header and service elements to build or amend the appropriate process scheme from possible interconnections. The production of the order starts during *operate* phase. On completion of the order the process scheme is terminated in the *terminate* phase (*e.g.*, in scenario III). During the *operate* phase, the process elements monitor plant conditions and invoke a new round of reconfiguration if a major failure or a disturbance is detected (*e.g.*, as in scenario IV where a unit element fails or in scenario VII where the demands for end-products change).

In what follows, we use the first five scenarios to describe the nature of interactions between process elements in all three cases in Table 7.3 individually. We assume that the above interaction sequence operates in background and focus only on the key interactions between unit and product elements and also the outcomes of these interactions in terms of the structure of resulting process schemes.

In the description we use the following notation: U followed by the number in process layout to refer to a unit element, PR followed by the product name to refer to a product element, and T followed by the number in Table 7.1 to refer to a task.

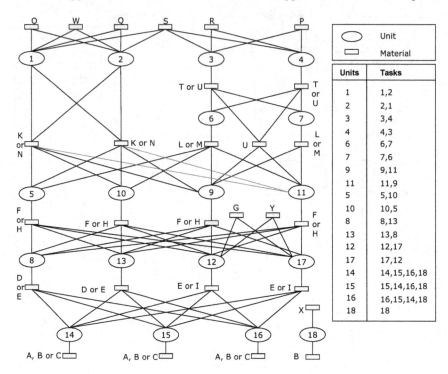

Fig. 7.6. Process layout: Case 3 - multiple tasks, flexible connectivity

Case 1: Single Task, Fixed Connectivity, Product-centric Approach

The first example demonstrates the interactions for Case 1 using the product-centric approach for recipe mapping in the define phase. We describe the operations required for scenarios I-V. As stated earlier, in a product-centric approach the product elements are supplied with product recipes shown in Fig. 7.2 and their role purely is to assign those tasks on the dark-lined sequence to suitable unit elements, but if this is not achievable an alternative sequence may be chosen.

I. *Start product B order:* On arrival of a new order for product B, a new product element PR$_B$ is created during identify phase. In the next define phase, PR$_B$ then engages with all unit elements in the process to perform recipe mapping. It announces each of its processing task on the preferred task sequence (*i.e.,* dark-lined sequence for product B in Fig. 7.2) and assigns them to suitable unit elements based on the responses received. If no response is received for any of the tasks, then it may choose a nearest alternative sequence in the recipe to minimise de-assignment of previously assigned tasks. The unit elements involved in the resulting tentative schemes then refine these schemes into a single scheme that is used for reconfiguration and operation. Fig. 7.7(a) shows the layout of the final scheme if this whole sequence is completed satisfactorily. Note that PRB only selects unit elements which match with its preferred sequence.

II. *Add product A order:* Next, a new product element PRA is created for product A which leads to a similar round of interactions involving all unit elements. Fig. 7.7(b) shows the process scheme resulting from these interactions. It now comprises both PRA and PRB. Note that although unit element U5 also produces material F for product A which is also an ingredient in producing B, it is in fact not involved in the production of B because the task T5 is not allowed on the preferred sequence of PRB. Unit element U1 is however involved in both schemes.

III. *Changeover from product B to C:* The changeover leads to creation of PRC and removal of PRB. Initially when PRC is created all unit elements in the process get involved in the creation of new process scheme for C, but as PRB is removed, those involved in producing B also terminate their tasks for PRB. Figs. 7.7(c) and 7.7(d) depict the resulting process schemes when all three product elements exist together and when only PRA and PRC remain.

IV. *Unit element U11 fails:* Assume unit element U11 of type 'reactor type 1' fails and cannot supply material H any more. Thus, unit elements U17 and hence U16 also cannot continue with their tasks. An alternative source of H is thus required. Since no other unit elements on the preferred task sequence of PRc can supply H, PRC invokes a new round of interactions. During identify phase, the requirement imposed for PRc is to choose an alternative sequence that can produce C. The subsequent interactions then follow as in previous scenarios. Fig. 7.7(e) shows the process scheme based on a different sequence involving U10.

V. *Unit element U11 Rejoins:* The incoming unit element in this case announces its capability to all product and unit elements. Since PRC can make use of its facility to revert back to its preferred sequence, it has a choice whether to continue with the ongoing scheme or to choose this preferred option. Assuming the decision rule defined in PRC is to choose the preferred sequence where possible, it invokes a new round of interactions and reassigns the tasks as appropriate to return to the scheme in Fig. 7.7(d).

Note that unit elements in the above description only possess localised knowledge of their task capability (*i.e.,* the type of unit operation they can perform). They do not have the knowledge of preferred task sequence or the materials or services associated with the individual tasks. Such information is

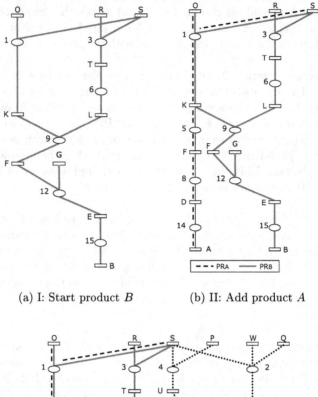

(a) I: Start product B (b) II: Add product A

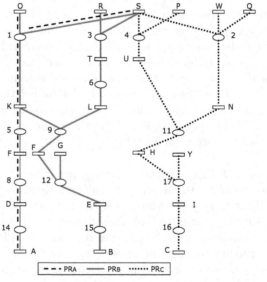

(c) III: Products A, B, C together

Fig. 7.7. Illustrations of process schemes: Case 1, Scenarios I, II, III

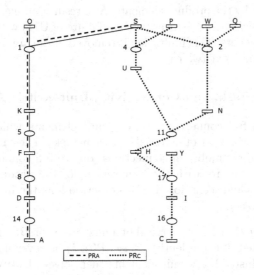

(d) III: Changeover from *B* to *C*

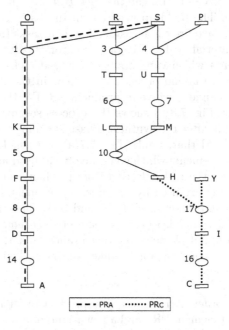

(e) IV: U11 fails

Fig. 7.7. Illustrations of process schemes: Case 1, Scenarios III, IV

assigned to them by the product elements. As a result unit elements are unable to respond to process disturbances (such as failure of U11) without interacting with product elements. This limitation is removed in the unit-centric approach as discussed in the next two cases.

Case 2: Single Task, Full Connectivity, Unit-centric Approach

Case 2 refers to full connectivity among unit elements which are now also defined with the exact set of materials for their tasks. Their role now involves finding from a large number of connections (due to full connectivity) a single process scheme that fits with the requirement of the product order. Below we describe the same scenarios (I-V) to describe how the interactions would proceed in this case.

I. *Start product B order:* On arrival of a new order for B, a product element PRB is created during identify phase. PRB is however not supplied with any recipe. Instead, the unit elements themselves identify the tasks they can to use to produce requested material. The interactions thus proceed in a demand-pull fashion starting from the end-product B. Since two unit elements, U15 and U18, can produce B, both initiate building a new process scheme. The build-up proceeds in the backward direction. Both units attempt to acquire the feedstock for their tasks (material E for T15 of U15 and material X for T18 of U18) from upstream unit elements. All unit elements which can supply these feedstocks get involved. U18 will however find that no other unit element in the process can supply X. It therefore cannot involve in producing B. For U15, the interactions proceed further. Fig. 7.8(a) shows the process scheme that results from these interactions after the synthesis phase is completed.

It can be noted that, unlike Fig. 7.7(a) in Case 1, the final scheme includes all unit elements which could involve in producing B as there are no constraints on the task selection from product recipe. Note also that materials L and U are used by multiple unit elements – material L used by U9 and U10 and material U by U7 and U11. These materials thus fall along two different branches of the same process scheme that lead to product B. Thus, if a unit element in either branch fails, the unit elements in the other branch should be able to take over its load within their capacity (see scenario IV).

II. *Add product A order:* Arrival of a new product order for A leads to creation of product element PRA and a new round of interactions. Since only U14 can produce A, it initiates the formation of a process scheme. The interactions proceed similar to previous scenario. Fig. 7.8(b) shows the final process scheme. Note that material F is now involved in both process schemes. Thus, when U8 makes its request for F, both U5 and U9, which are already engaged in producing B, also engage in producing A. These elements subsequently reallocate their feedstock demands for K and L by re-interacting with upstream supplier unit elements. The effects of this reallocation incrementally propagates to other unit elements in both process schemes.

III. *Changeover from product B to C:* Changeover from product B to C leads to creation of PRc and removal of PRb. Fig. 7.8(c) and 7.8(d) show the process schemes when all three product elements coexist and when only PRa and PRc remain. In total, two new unit elements U16 and U17 get involved while U12, U13 and U15 are removed. Subsequently, the latter three elements also terminate their interactions for material E and then for materials F and H. The unit elements upstream reallocate their material demands accordingly.

IV. *Unit Element U11 Fails:* Failure of U11 leads to an abrupt termination of all its interactions with upstream and downstream unit elements. U2 will thus also be removed from process scheme for product C. Since U10 also supplies material H, it takes over the load from U11 within its capacity and reallocates its feedstock demands as appropriate. This response to failure emerges directly by the interactions between failed element U11, and the affected elements U17 and U10. The resulting scheme from these interactions is not shown here for brevity, but its structure can be easily derived from Fig. 7.8(d).

V. *Add a New Unit U19 or U11 Rejoins:* In either case the incoming unit element announces its presence to other unit elements in terms of the materials it can supply. Unit elements which can use this facility, *e.g.*, U17 here, then interact with it to reallocate its feedstock demands accordingly. The interactions should thus reinstate the scheme in Fig. 7.8(d).

It can be seen that by supplying material-specific information for their tasks, the unit elements are able to manage recipe mapping activity in a distributed manner. This distribution helps manage a change or failure in a graceful manner compared to product-centric approach in Case 1. More importantly, the unit elements are also capable of selecting processing tasks that are known locally that otherwise may not be specified by the developers of product recipes situated often remotely. A benefit of this can be seen by comparing Fig. 7.7(a) with Fig. 7.8(a). In the former only those unit elements whose tasks match with the recipe are selected, while in the latter all unit elements which can involve in making B are selected. The latter is thus also likely to have a better chance to respond to a change or failure than the former.

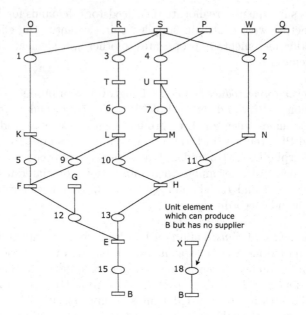

(a) I: Start product B order

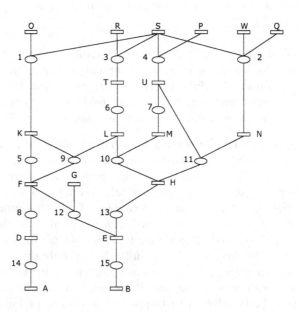

(b) II: Add product A order

Fig. 7.8. Illustration of process schemes: Case 2, Scenarios I, II

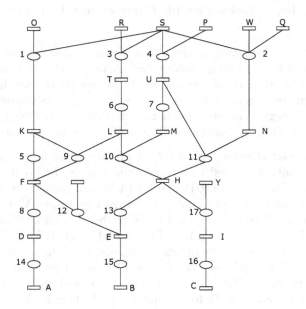

(c) III: Products A, B and C together

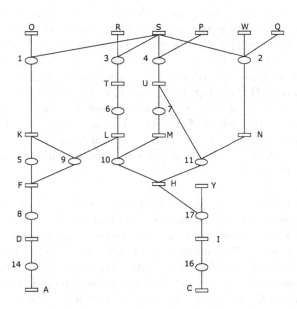

(d) III: Changeover from product B to C

Fig. 7.8. Illustration of process schemes: Case 2, Scenarios III, IV

Case 3: Multiple Tasks, Flexible Connectivity, Unit-Centric Approach

Case 3 (Fig. 7.6) removes the limitation of single task capability and also considers flexible connectivity among unit elements. The unit elements now receive a combined choice of selecting local tasks and/or the supplier elements for their feedstocks in developing the process schemes. As the description of scenarios I-V next illustrate, this flexibility leads to an added freedom in responding to emerging changes or disturbances than the previous two cases.

I. *Start product B order:* On arrival of product order for B, a new order PRB is similarly created with a new round of demand-pull interactions. Four unit elements, U14, U15, U16 and U18, can produce B. For U18 the scheme cannot proceed as no unit element can supply X. For the other three elements the interactions proceed further. U8, U13, U12 and U17 can supply E: U8 and U13 via task T13 and U12 and U17 via T12. All four hence get involved and try to extend the process scheme by acquiring their feedstock H for T13, and $\{F, G\}$ for T12. Note that there are multiple task sequences available for producing both F and H. Unit elements U5 and U10 can use task T5 to produce F and T10 for H. Similarly U9 and U11 can use task T9 for F and T11 for H. These unit elements thus face a choice when they attempt to acquire feedstocks for these alternative tasks. The final selection of a specific task would occur during synthesis phase when unit elements refine these tentative process schemes into a single scheme. To simplify the discussion, we consider an assumption that the unit elements select the first task in the sequence in Fig. 7.6 when they have such a choice, *i.e.*, U5 and U10 select tasks T5 and T10 respectively, while U9 and U11 select tasks T9 and T11. The complete process scheme thus developed involves six more unit elements from upstream. We do not show the resulting scheme for brevity. We can see however that compared to Cases 1 or 2, the introduction of multipurpose character of unit elements with flexible connections leads to increased choice available to unit elements for producing B.

II. *Add product A:* On arrival of product A order a new product element PRA is created which announces its requirements. Unit elements U14, U15 and U16 which all can produce A are however engaged with PRB. These unit elements, while continuing with their tasks, initiate a new round of interactions to develop the tentative schemes for A. During synthesis phase these elements then use a production goal to decide whether or not to de-commit from their existing tasks and involve in the production of A. For simplicity of discussion, we assume that U14 prefers the first task (*i.e.*, T14) in sequence in Fig. 7.6 over T15. It hence de-commits from T15. Its capacity for producing B is transferred to U15 and U16 as appropriate. The same decision rule also extends to unit element U8 which de-commits from its task to involve in the process scheme for A.

III. *Changeover from product B to C:* Creation of PRc leads to U16 changing its task from T15 to T16 (by continuing with the assumption of task preference). Unit element U15 is now the only element producing B. The same change also occurs for U17 which changes from T12 to T17. Subsequently, when PRB is removed, all unit elements which were engaged with product B de-commit from their tasks. These unit elements now become available and announce their capabilities to other unit elements. The process schemes for products A and C are thus revised to involve these unit elements as appropriate.

IV. *Unit Element U11 fails:* The failure of U11 results in a partial loss of supply for material H. U10 which is also involved in producing H attempts to take over its load through T10. It is possible here that U9 could replace U11 if material H is more essential than F when comparing the importance of product C to product A or if the capacity for F can be shifted to other unit elements in the process scheme for A. The unit elements make these decisions during synthesis phase in deciding the final process scheme and their local operating settings.

V. *Add a New Unit U19 or U11 Rejoins:* The new element announces its capabilities and gets involved in the interactions. If U9 has changed its task in the previous scenario than it has a choice to revert back to its original task since an alternative supplier of H is available. Using the rule of task preference, it will do so. The outcome of the interactions should lead to reinstating the scheme in Scenario III.

As we can see, the enhancement in local capabilities of unit elements aided by the flexibility in their interconnections leads to an increased choice and re-configurability in all scenarios described above compared to Cases 1 and 2. This observation thus provides a crucial guideline in organising the process elements based on the distributed architecture and the interaction model discussed in previous chapters.

7.4.4 Coordination of Distributed Operating Settings

The coordination of local and network parameters of process elements occurs via their distributed interactions. During synthesis of a process scheme from multiple tentative schemes this coordination involves various mixed-integer decisions such as which tasks and hence supplier elements should be selected (as discussed in Case 3). While a complete computational framework covering all such decisions is beyond the scope of this text, the algorithm presented in Chapter 6 provides a sensible framework to define these interactions in a mathematical form.

Below, we illustrate the developments in Chapter 6 by applying them to the current example, in particular to scenario VI. We use the layout in

Fig. 7.8(c) which includes all 17 unit elements involved in making A, B, and C. It is assumed that the unit elements have found this process scheme during synthesis phase and the aim of distributed algorithm is to find the settings of local and interaction parameters.

Modelling the Local Dynamics of Process Elements

The local dynamics and interconnections of unit elements are modelled using the linear, dynamical model presented in Section 6.2. In particular, the demands for outgoing products of a unit element are treated as disturbances and the demands that it places for materials and services to upstream unit elements and service elements are treated as the manipulated variables that it controls. Tables C.9 and C.10 in Appendix C define the problem data for all 17 units in Fig. 7.8(c). The problem data is implemented using the framework defined in Appendix B.

Operation of the Distributed Algorithm

Each of the 17 unit elements is supplied with the generic unit software module defined in Appendix B. As stated therein, the module is generic in that it applies to all four types of junction block connections of a unit element. Depending on the type of junction block generated in the synthesis phase, the optimality cuts generated are varied as appropriate.

To model scenario VI, we assume that all three products initially have a demand of 10 deviation units from a nominal set-point. Tables C.11 and C.12 in Appendix C summarise the progress of the distributed algorithm for local state and manipulated variables $x_{i,z}$ and $u_{i,z}$ for unit element $i = 1, \ldots, 17$, where z refers to z^{th} element of x_i and u_i for unit element i. Note that we have numbered unit elements by $i = 1, \ldots, 17$ instead of the ordering (n, s) in Chapter 6. It can be seen that the algorithm converges to optimal solution within three or four iterations, although an accuracy of four digits requires further iterations.

As a next step to scenario VI, the demand for all three products is changed from 10 to 20 deviation units after iteration 10. Fig. 7.9 shows the effects of approximate cut updates on the sub-problems of unit elements 14 and 16. Fig. 7.10 demonstrates the effects of this change in demands in terms of the feedstock demands that unit elements 14 and 16 place to their upstream units. The solution algorithm is able to absorb this change and converge to a new optimum solution after 21 iterations.

Fig. 7.11(a) on Page 148 shows the computational performance of the distributed algorithm in terms of floating-point calculations (flops) required as compared to a centralised algorithm for solving a series of 30 different data-sets for the same problem. In the case of the distributed algorithm, we terminate the algorithm if the number of iterations reaches more than 20. We can observe that the distributed algorithm, although not as efficient (which is

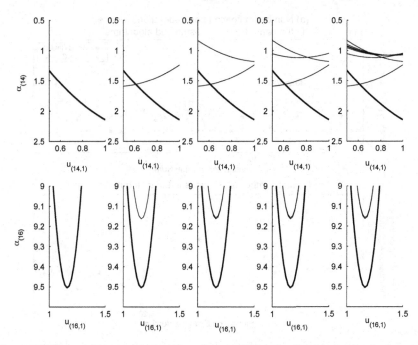

Fig. 7.9. Effects of approximate cut updates on the value functions of unit elements 14 and 16 sub-problems

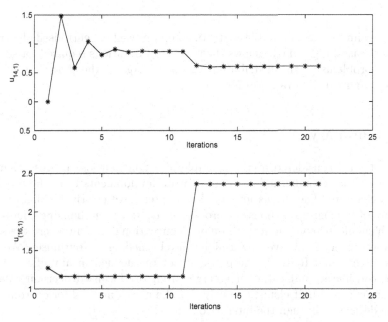

Fig. 7.10. Effects of change in terminal demands of unit elements 14 and 16

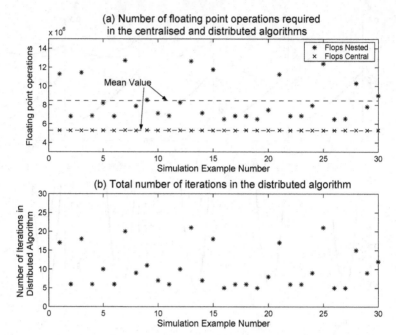

Fig. 7.11. (a) Comparison of floating -point operations (flops) required between centralised and distributed algorithms; (b) Number of iterations required in distributed algorithm, both for 30 data-sets

not the aim for reconfigurable control), compares well to centralised algorithm in most cases. Fig. 7.11(b) shows the number of iterations required for solving all 30 problems in the distributed algorithm. Again, the iterations remain limited and within a range of 5 to 20.

7.5 Summary

This chapter considered a case example of a multipurpose process plant to illustrate the reconfigurable process control developments from the previous three chapters. The discussions in the chapter have clearly highlighted the nature of bottom-up response of process elements under changing conditions which should be compared to those of a conventional system where the same response would be derived by a higher-level scheduler or optimiser. We emphasise that in DRPC this response is not pre-defined in any of the three cases considered, but rather it emerges from the localised decision rules of the product and unit elements involved and a global method for coordinating these decisions through the interaction model.

8
Conclusions

The research in this monograph has presented a distributed approach to the control of process operations that require a high degree of reconfigurability. A distributed coordination approach based on the distributed paradigms of holonic manufacturing and supply chain management was developed in this text to develop a blueprint for reconfigurable process control systems.

In this chapter we now summarise the key contributions from this work, the limitations of the research, and the areas for future work where this research can be extended.

8.1 Main Contributions

The DRPC approach promotes a bottom-up method to the design and integration of process control systems. The bottom-up approach is preferred as it enables rapid integration and reconfiguration, both during and after the design life-cycle. The overall design method of a DRPC system then operates in the following sequence:

 i. a top-down decomposition of the top-level requirements into corresponding low-levels requirements;
 ii. interpretation of the low-level requirements into the selection of process elements and assignment of their control responsibilities; and
 iii. design, implementation, and integration of process elements into a complete system in a bottom-up manner.

While the top-level requirements in step (i) are expected to cover a range of possible production scenarios, they may not – and need not be – exhaustive as the design should be reconfigurable enough to allow for new requirements at any stage in the design or operation life-cycle. Similarly, the decomposition of top-level requirements may only be required to the level of abstraction where they can be delivered by the self-contained design of process elements.

For example, in case of a reaction operation, if the design of the reactor element permits, the low-level requirement assigned to that element should be of the form ⟨perform reaction 'X'⟩ as opposed to specifying in detail the operation of individual actuators or the control policies. In steps (ii) and (iii), it remains important that the principle of *low and least commitment* (Valckenaers & van Brussel 2005) is employed so as to induce a maximum level of flexibility between elements to operate over a range of conditions, both planned or unplanned.

Within this framework, the developments in Chapters 4, 5 and 6 provide the concepts and necessary guidelines to develop a design process from which an RPC system can be developed. These developments can be summarised as below.

Distributed Control Architecture

The description in Chapter 4 started with introducing a new concept of *process element* as a stand-alone, modular building block of a DRPC system. It was suggested that the identification of process elements is done based on their physical involvement in the process while also observing that each element must have at least one but possibly more decisions that it can regulate on its own. A systematic method to perform such an identification was developed by Bussmann (Bussmann, Jennings & Wooldridge 2001, Jennings & Bussmann 2003, Bussmann 2003) in the context of a discrete process.

Chapter 4 later also identified and defined the structure of four key process element types as forming any DRPC system. When engineering a DRPC system, this identification of element types should be used to develop a library of multipurpose designs of process elements that can be deployed as 'off-the-shelf'. In step (ii) in the overall design method discussed at the beginning of the section, an appropriate element can be selected from this library to meet the low-level requirements.

Distributed Interaction Model

Chapter 5 introduced a distributed model for managing the interactions between process elements. The proposed model builds upon two key aspects: (a) the supplier-customer design of process elements is used based on which these elements acquire their feedstocks and services from respective supplier elements and (b) the demand-pull type interaction behaviour is used to build the plantwide process schemes. The five steps of the reconfiguration process (Fig. 5.1) then define the sequence of interactions that elements must follow to develop or reconfigure a process scheme in response to changing plant conditions. The interaction model also characterised product-centric and unit-centric approaches for recipe mapping as the two distinct approaches, the former being more appropriate for high variety of products while the latter more appropriate for frequent changes between the same products.

Distributed Coordination Algorithm

One particular aspect of the interaction model – that of identifying the local operating settings of unit elements – was considered in Chapter 6 as an example of developing a distributed coordination strategy for managing the localised operations of process elements. It was shown that an economic interpretation of so-called *nested decomposition* algorithm can provide a systematic method to implement the intended demand-pull type, price-demand guided interactions between elements in a mathematical form. While the classical techniques in nested decomposition (Ho & Manne 1974, O'Neill 1976, Wittrock 1985) apply generally to series-connected networks, their extension using so-called *approximate cut update* technique enables their use in process networks of arbitrary but acyclic nature. The implementation of the algorithm is centered around a single, general-purpose unit module (see Appendix B) that covers all possible combinations of connections in which a unit element can be located within an acyclic network. For a practitioner intending to deploy the algorithm (even as part of conventional hierarchical model), it thus suffices to develop a single such module that can be incorporated as part of the design of any unit element.

8.2 Limitations of the Research

This work forms one of the first attempts at using distributed coordination for developing reconfigurable process operations. Since the scope of the development is wide, naturally some limitations remain. Below we describe key such limitations where further work could be useful.

- *Organisation of process elements:* The control architecture in Chapter 4 was developed to the extent of identifying the types of process elements and defining their structure, *i.e.,* data models, control functions and connections. An account of the flexibility in terms of the alternative configurations in which they can be organised (when developing a reconfigurable process plant from grass route) or the existing configurations can be changed (when revamping an existing control system) was not discussed at length as this forms a question of practical implementation. Some examples of alternative choices were given in Chapter 7, however for more specific design practices, it will be necessary to quantify how much flexibility is sufficient and cost-effective so as to address the end-user needs for a foreseeable future.
- *Interaction behaviour of header and service elements:* The discussions in Chapters 5 and 6 can be extended to cover the interactions of header and service elements. For header elements, the issue is to define systematic methods for reconfiguring the process routes in a coordinated manner especially when the transients occur. For service elements, the methods for

service distribution must be linked closely with the material exchange interactions because the consumer elements which use these services may also be connected via physical routes and therefore change in service allocation at one point can affect the operations in other parts of the network.

- *Nonlinear dynamics, transients and recyles in distributed coordination:* The distributed control problem used as a basis in Chapter 6 was limited to a simplified problem based on linear, steady-state dynamics model. The solution strategy developed therein could be generalised to other class of problems involving non-linear and/or dynamical models – the former class of problems can arise in distributing the optimisation layer while the latter in distributing the advanced control layer.

8.3 Future Challenges

In addition to the above limitations, there remain other broader issues concerned with this research where further challenges remain.

- *Tools for human interactions:* Tools for human interactions should be developed to define where and how a human role is involved as part of the reconfiguration process. Inclusion of the human role will be important in DRPC in identifying that a reconfiguration is necessary, and defining a feasible configuration from the available choices. It is envisaged that the actual reorganisation of elements will be automated in future, but humans will play a key role in making this happen.
- *Design methodology and integration within industrial practice:* A comprehensive design method for DRPC should be developed that offer complete guidelines for developing a system that fulfills end-user requirements for a sufficiently foreseeable future. The method should be also detailed enough to enable an engineer not familiar with distributed concepts to perform design tasks with little or no external help. For the short-term future, the design method should also consider a migration strategy for operators of the existing plants to incrementally move towards building the DRPC system envisioned here. Some aspects of migration have been discussed in Section 4.3.
- *Management of virtual enterprises:* Finally, the most closely related field to the research in this text is the field of managing virtual enterprises themselves. In the past, supply chain management has benefited from control engineering tools, *e.g.*, for studying inventory control (Deonckheere, Disney, Lambrecht & Towill 2002, Perea-López, Grossmann, Ydstie & Tahmassebi 2000, Chaib-draa & Müller 2006). The distributed coordination strategy developed in Chapter 6 could be extended to solve the large-scale control problems associated with managing a virtual enterprise in a distributed manner.

A

Appendix to Chapter 6: Background Concepts

This appendix discusses the background concepts for Chapter 6. The so-called *basic sensitivity theorem* from sensitivity analysis is described in the next section to characterise the parametric sensitivity of an optimisation problem to variations in an exogenous parameter vector forming an internal part of the problem. The description here is adapted from Fiacco (1983). Section A.2 then explains the concept of primal decomposition in a greater detail. The discussion therein is based on Geoffrion (1970) and Grothey *et al.* (1999).

A.1 Basic Sensitivity Theorem

Consider the following optimisation problem

$$
P_{rhs}(\theta) \quad
\begin{cases}
\underset{x(\theta)}{\text{minimise}} \ f(x) \\
\quad \text{s.t.} \quad h_i(x) = \theta_i, \quad i = 1, \ldots, p, \\
\qquad\qquad g_j(x) \le \theta_{p+j}, \quad j = 1, \ldots, m.
\end{cases}
\tag{A.1}
$$

where $x \in \mathbb{R}^n$ and θ is an exogenous parameter vector in $\mathbb{R}^k$.

Let x^* be the strict local optima of $P_{rhs}(\theta)$, and λ^* and μ^* be the corresponding Lagrange multipliers for equality and inequality constraints when $\theta = 0$ (*i.e.*, there exists a neighborhood of x^* where there does not exists any feasible $x \ne x^*$ such that $f(x,0) \le f(x^*,0)$). Let $f^*(\theta) \equiv f(x(\theta))$ denote the (local) optimal value function of $P_{rhs}(\theta)$ when $x(\theta)$ solves $P_{rhs}(\theta)$ for θ near 0.

The following result characterises the sensitivity of $f^*(\theta)$ for variations in θ in a neighborhood of 0.

Theorem A.1 (Optimal Value Function Derivatives for $P_{rhs}(\theta)$, Corollary (3.4.4) in (Fiacco 1983)).
   If

(i). *the functions defining $P_{rhs}(\theta)$ in a neighborhood of $\theta = 0$ are twice continuously differentiable in x in a neighborhood of x^*,*

(ii). *the second-order sufficient conditions for a local minimum of $P_{rhs}(0)$ hold at x^*, with associated Lagrange multipliers λ^* and μ^*,*

(iii). *the gradients $\nabla g_i(x^*, 0)$ (for i such that $g_i(x^*, 0) = 0$) and $\nabla h_j(x^*, 0)$ (all j) are linearly independent, i.e., Slater constraint qualification is satisfied (see Bazaraa, Sherali & Shetty (1993) for a definition of Slater condition),*

(iv). *$\mu_i^* > 0$ when $g_i(x^*, 0) = 0, i = 1, \ldots, m$, i.e., strict complementarity slackness holds, then*

then in a neighborhood of $\theta = 0$,

(i). $\nabla_\theta f^*(\theta)^T = \begin{bmatrix} \lambda(\theta) \\ \mu(\theta) \end{bmatrix}$, *and*

(ii). $\nabla_\theta^2 f^*(\theta) = \begin{bmatrix} \nabla_\theta \lambda(\theta) \\ \nabla_\theta \mu(\theta) \end{bmatrix}.$

□

The above result, also known as *Lagrange Multiplier Sensitivity* theorem, suggests that for any given θ, the Lagrange multipliers resulting from solving $P_{rhs}(\theta)$ provide the sensitivity of the optimal value function for variations in θ in a close neighbourhood of 0. In economic sense, these multipliers can also be interpreted as the so-called *shadow prices* or *marginal costs* defining the variation in optimum supply costs for a unit change in the product demands θ (Edgar, Himmelblau & Lasdon 2001). The theorem was used in Section 6.5.2 to develop the so-called *approximate cut update technique* for the parametric two-units problem.

A.2 The Concept of Primal Decomposition

This section now describes the concept of so-called *primal decomposition* from (Geoffrion 1970) for the types of problems considered in Chapter 6.
   Consider the following optimisation problem for a large-scale system comprising N subsystems.

Problem A.2.

$$\underset{x}{\text{minimise}} \sum_{i=1}^{N} f_i(x_i)$$

$$\text{s.t.} \quad \sum_{i=1}^{N} r_i(x_i) \leq R, \quad i = 1, \ldots, N$$

$$x_i \in X_i.$$

where x_i is an n_i dimensional vector of real-valued variables associated with ith subsystem, X_i is a subset of $\mathbb{R}^{n_i}$, f_i is the cost function for ith subsystem. The constraint $\sum_{i=1}^{N} r_i(x_i) \leq R$ $i = 1, \ldots, N$ connects all subsystems and may represent a limited quota of resources being shared by the subsystems.□

The presence of this linking constraint means that the local sub-problems cannot be solved independently. However, by assigning individual local limits $\hat{r}_i,\ i = 1, \ldots, N$, the problem can be recast to the following separable form:

$$\underset{\hat{r}}{\text{minimise}} \ \sum_{i=1}^{N} \alpha_i(\hat{r}_i) \ \text{ subject to } \ \sum_{i=1}^{N} \hat{r}_i \leq R, \tag{A.2}$$

where $\alpha_i(\hat{r}_i)$ are the *value functions* defined as: for all $i = 1, \ldots, N$:

$$\alpha_i(\hat{r}_i) \triangleq \underset{x_i, u_i}{\text{minimise}} \ f_i(x_i, u_i) \tag{A.3}$$
$$\text{s.t.} \quad x_i = h_i(x_i, u_i)$$
$$x_i \in X_i, u_i \in U_i$$
$$r_i(x_i, u_i) \leq \hat{r}_i.$$

A primal decomposition algorithm (Geoffrion 1970) exploits this restructured form to solve the problem via iterative, two-level coordination mechanism. The problem in Eq. A.2, denoted as SP_M, is assigned to a higher-level coordinator, and the sub-problems in Eq. A.3, denoted as SP_i, are assigned to individual subsystems $i = 1, \ldots, N$. The two-level coordination scheme then operates as shown in Fig. A.1.

The coordinator first solves its problem SP_M to find a feasible value of $\hat{r}_i$ for all $i = 1, \ldots, N$ such that $\sum_{i=1}^{N} r_i(x_i) \leq R$, $i = 1, \ldots, N$. These $\hat{r}_i$ are then sent to the subsystems. For a given value of $\hat{r}_i$, the subsystems then solve their problems SP_i to find the optimum values x_i^* and f_i^*, and the optimal Lagrange multipliers λ_i^*, which are returned to the coordinator. Using all responses, the coordinator problem is revised using to find a new value of $\hat{r}_i$ that can reduce the total cost $\sum_{i=1}^{N} \alpha_i^*$ for all subsystems. The interactions between two levels thus repeat until $\hat{r}_i, i = 1, \ldots, N$ converge to a fixed point.

Note that to decide optimal distribution $\hat{r}_i$, the coordinator needs an explicit representation of the value functions α_i. Similarly, it also needs to ensure that the value of $\hat{r}_i$ passed to subsystem i is such that the sub-problem SP_i in Eq. A.3 remains feasible, *i.e.*, $\hat{r}_i$ belongs to a set

$$V_i \triangleq \{ \ \hat{r}_i | \ \exists \ x_i, u_i \text{ for which } x_i = h_i(x_i, u_i) \tag{A.4}$$
$$r_i(x_i, u_i) \leq \hat{r}_i$$
$$x_i \in X_i, u_i \in U_i \}.$$

In (Geoffrion 1970), Geoffrion suggested a *tangential approximation* scheme to build an approximation of both α_i and V_i in an iterative manner and use that in solving the problem in a decomposed form.

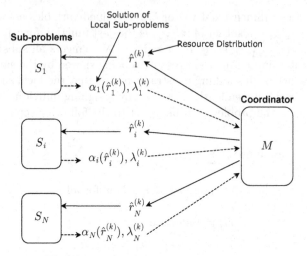

Fig. A.1. Primal decomposition scheme

A.2.1 Approximation of $\alpha_i(\hat{r}_i)$

Under a set of suitable regularity assumptions (see assumptions A1 and A2 in Geoffrion 1970), the value function $\alpha_i(\hat{r}_i)$, for any $i = 1, \ldots, N$ can be represented as a supremum of its *linear supports*, where a linear support has the following definition.

Definition A.3 (Linear Support (Geoffrion 1970, Rockafellar 1970)). *A linear support to a smooth, convex function $f(x)$ is any linear function whose value is always less than or equal to that of the $f(x)$, with equality holding at some point in the domain of $f(x)$, i.e., in the range of x values where $f(x)$ is finite (see Fig. A.2). A linear support can be represented as the inequality*

$$f(x) \geq f(\bar{x}) + \nabla f(\bar{x})(x - \bar{x})$$

where $\nabla f(\bar{x})$ denotes the gradient of $f(x)$ with respect to x at $\bar{x}$. □

We then have the following result.

Proposition A.4 (Theorem 3, Geoffrion (1970)). *Let $\bar{\hat{r}}_i$ be such that the problem in Eq. A.3 has an optimal solution $(\hat{x}_i, \hat{u}_i)$ and an optimal multiplier vector $\lambda_i(\bar{\hat{r}}_i) \in \Lambda_i \subseteq \mathbb{R}^{m_{r_i}}$ associated with $r_i(x_i, u_i) \leq \bar{\hat{r}}_i$ constraints. Then the function*

$$\alpha_i(\bar{\hat{r}}_i) - \lambda_i^T(\bar{\hat{r}}_i)(\hat{r}_i - \bar{\hat{r}}_i), \tag{A.5}$$

is a linear support to $\alpha_i(\hat{r}_i)$ at $\bar{\hat{r}}_i$, and that

$$\alpha_i(\hat{r}_i) \geq \alpha_i(\bar{\hat{r}}_i) - \lambda_i^T(\bar{\hat{r}}_i)(\hat{r}_i - \bar{\hat{r}}_i) \quad \text{for all } \hat{r}_i$$

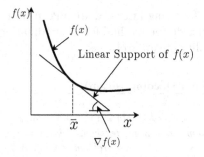

Fig. A.2. Linear support

Proof. From the basic sensitivity theorem (see (Fiacco 1983) or Theorem (A.1) in this appendix), it is known that

$$\nabla \alpha_i \equiv -\lambda_i(\hat{r}_i).$$

The statement then follows from the definition of linear support. □

This result can be used to develop an iterative algorithm that incrementally builds an improving approximation of α_i using its linear supports. For example, consider a case where we are at the Kth iteration in an algorithm for solving Prob. A.2 and that an optimal value of the multiplier vector $\lambda_i^{(k)}$ for all $i = 1, \ldots, N$ has been found at each point $\hat{r}_i^{(k)}$, $k \in \mathbf{K}$, where k indexes the iteration count and $\mathbf{K} \equiv \{1, \ldots, K\}$, then the corresponding approximation to α_i at iteration K can be written as a piecewise linear function.

$$\alpha_i^{(K)}(\hat{r}_i) \triangleq \sup_{k=1,\ldots,K} \left\{ \alpha_i(r_i^{(k)}) - (\lambda_i^{(k)})^T(\hat{r}_i - \hat{r}_i^{(k)}) \right\}, \qquad (A.6)$$

Fig. A.3 depicts the nature of this approximation. The figure on the right therein shows how the linear supports approximate the actual value function $\alpha_i(\hat{r}_i)$ in the figure on the left.

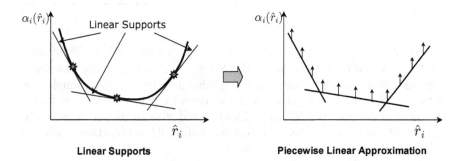

Linear Supports   **Piecewise Linear Approximation**

Fig. A.3. Piecewise linear approximation of $\alpha_i(\hat{r}_i)$

We now have the following intermediate algorithm for solving Prob. A.2, where we are yet to decide the method for approximating the feasible regions V_i that ensure feasibility of sub-problems.

Algorithm A.1 (Primal Decomposition)

Step 0: Initialise: Take a feasible solution $\hat{r}_i^{(0)} \in V_i$, for all $i = 1, \ldots, N$ in Eq. A.2 such that the sub-problems in A.3 are feasible. Take $K = 1$.

Step 1: Sub-problem SP_i : Solve sub-problems in Eq. A.3 by any suitable optimisation solver. Recover an optimal multiplier vector $\lambda_i^{(K)}$, $i = 1, \ldots, N$.

Step 2: Master Problem SP_M: Add a new linear support to the master problem using A.5. Solve the resulting tangential approximation of the master problem using a suitable optimisation solver to get a new $\hat{r}$. This can be written as,

$$
\begin{aligned}
\underset{\hat{r}}{\text{minimise}} \quad & \sum_{i=1}^{N} \alpha_i^{(K)}(\hat{r}_i) \\
\text{s.t.} \quad & \sum_{i=1}^{N} \hat{r}_i \leq R, \\
& \hat{r}_i \in V_i, \quad i = 1, \ldots, N.
\end{aligned} \tag{A.7}
$$

By exploiting the form of piecewise linear approximation $\alpha_i^{(K)}(\hat{r}_i)$ as in Fig. A.3, the above problem can be written more conveniently as,

$$
\begin{aligned}
\underset{\hat{r}, \nu_i}{\text{minimise}} \quad & \sum_{i=1}^{N} \nu_i \\
\text{s.t.} \quad & \nu_i \geq \alpha_i(r_i^{(0)}) - (\lambda_i^{(1)})^T(\hat{r}_i - \hat{r}_i^{(0)}), \quad i = 1, \ldots, N \\
& \vdots \\
& \nu_i \geq \alpha_i(r_i^{(K-1)}) - (\lambda_i^{(K)})^T(\hat{r}_i - \hat{r}_i^{(K-1)}), \quad i = 1, \ldots, N \\
& \hat{r}_i \in V_i, \quad i = 1, \ldots, N.
\end{aligned} \tag{A.8}
$$

where $[]^T$ denotes the transpose. Let the optimal solution of Eq. A.8 be denoted by $\hat{r}^{(K)}$.

Step 3: Iterate/Terminate: If $\hat{r}^{(K)}$ in Eq. A.2 is converged to a fixed point (see below), then terminate. Else, set $K := K + 1$ and return to step 1. □

Fig. A.1 outlines the information exchange between sub-problems SP_i, $i = 1, \ldots, N$ and master problem SP_M. Since $\alpha_i^{(K)}$ – being a piece-wise linear approximation – never overestimates the value function α_i, the optimal value of Prob. A.8 at each iteration provides a monotonically increasing lower bound on the optimal value of Prob. A.2. This coupled with the upper bounds that can be obtained from summing the optimal sub-problem objectives leads to a termination criteria for step K: Stop if

$$\sum_{i=1}^{N} \alpha_i^{(K)}(\hat{r}_i) - \sup_{0 \leq k \leq K-1} \{\sum_{i=1}^{N} \alpha_i(\hat{r}_i^{(K-1)})\} \leq \epsilon \qquad (A.9)$$

where ϵ is a small positive number. The first term represents the sum of optimal costs supplied by sub-problem solutions, while the second is the optimal cost obtained by the master problem using approximation of $\sum_{i=1}^{N} \alpha_i$ built over the last K iterations.

Remark A.5. Note that, at each iteration, we add N new linear supports in the master problem in Eq. A.8. That means, at the Kth iteration, Prob. A.8 will have KN linear constraints. This number can grow large as iterations proceed. We can instead interchange the summation $\sum_{i=1}^{N} \nu_i$ and the supremum of ν_i over its linear supports in Eq. A.8 to obtain an alternative master problem written as follows.

$$\underset{\hat{r},\nu}{\text{minimise}} \quad \nu$$

$$\text{s.t.} \quad \nu \geq \sum_{i=1}^{N} \left\{ \alpha_i(r_i^{(0)}) - (\lambda_i^{(1)})^T(\hat{r}_i - \hat{r}_i^{(0)}) \right\},$$

$$\vdots \qquad\qquad\qquad\qquad\qquad\qquad (A.10)$$

$$\nu \geq \sum_{i=1}^{N} \left\{ \alpha_i(r_i^{(K-1)}) - (\lambda_i^{(K)})^T(\hat{r}_i - \hat{r}_i^{(K-1)}) \right\},$$

$$\hat{r}_i \in V_i, \quad i = 1, \ldots, N.$$

Prob. A.10 now contains only K linear constraints instead of KN previously. $\square$

Remark A.6 (Optimality Cut). In light of (Benders 1962, Geoffrion 1970), the linear support in Eq. A.5 is also often referred to as *optimality cut*. This is due to the fact that each such new linear support improves the approximation of the value function α_i in the master problem and hence leads to an improved lower bound of the optimal global objective cost. $\square$

A.2.2 Approximation of V_i

A similar iterative scheme can be used to build an approximation of the feasible region V_i in Eq. A.4. The procedure starts with an initial approximation of $V_i^{(0)}$, for instance, a simple box constraint. At any iteration K of the Algorithm A.1, a value $\hat{r}_i \in V_i^{(K-1)}$ for all $i = 1, \ldots, N$ can then be tested to be feasible by checking the feasibility of sub-problem SP_i in Eq. A.3.[1] If feasible, we generate a new approximation $\alpha_i^{(K)}$ as before. Else, we exclude the

---

[1] Most commercial optimisation tools, such as the optimisation toolbox in MATLAB, provide a facility to check whether a problem is feasible or not before solving it.

respective region of $V_i^{(K-1)}$ by including in its approximation at the previous iteration a separating hyperplane that passes through the current value of $\hat{r}_i$. The formulation of this hyperplane can be obtained as follows.

A point $\hat{r}_i$ is in V_i if and only if it satisfies the system of linear constraints

$$\underset{u_i \in U_i}{\text{minimise}} \quad \lambda_i^T r_i(x_i, u_i) \leq \lambda_i^T \hat{r}_i \quad \forall \lambda_i \in \Lambda_i \tag{A.11}$$

where $\Lambda_i \triangleq \{\lambda_i \in \mathbb{R}_+^{m_i} : \lambda_i \geq 0 \text{ and } \sum_{j=1}^m \lambda_{ij} = 1\}$, where λ_i are the Lagrange multipliers obtained by solving the problem in Eq. A.11.[2] Any constraint of the form A.11 is a separating half-space for V_i. Since scaling of λ_i^T does not affect the solution of the above problem, a simple formulation of the separating hyperplane would be to include the constraint

$$\lambda_i^{(K)^T} r_i(x_i, u_i) \leq \lambda_i^{(K)^T} \hat{r}_i^{(K-1)} \tag{A.12}$$

into master problem SP_M where K denotes the iteration count in Algorithm A.1. This then allows for building an increasingly better approximation of V_i for any infeasible value of $\hat{r}_i$. Fig. A.4 shows the logic behind building V_i in this manner where two different supports to V_i are obtained at two different infeasible values of $\hat{r}_i$.

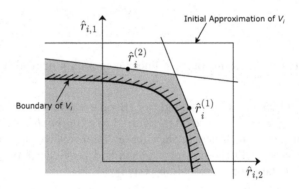

Fig. A.4. Approximation of V_i using supporting hyperplanes

Remark A.7 (Feasibility Cut). Similar to the naming of *optimality cut*, the hyperplane in Eq. A.11 is also referred to as *feasibility cut* (Geoffrion 1970). □

The above scheme of approximating V_i can now be incorporated in Algorithm A.1 to iteratively approximate both f_i and V_i together. In a worst-case

---

[2] Again, many commercial tools, such as the optimisation toolbox in MATLAB, automatically return these infeasibility related Lagrange multipliers, when the original sub-problem in Eq. A.3 is infeasible.

scenario, however, such an algorithm can lead to a situation where it requires an infinite number of iterations to approximate V_i, *i.e.*, it requires infinite separating hyperplanes to characterise V_i completely.

A useful remedy to this effect was suggested by Geoffrion (1972) which is that any such infinite sequence can be detected at an early stage through providing separate provisions in the algorithm, and can be terminated prematurely by extrapolating the sequence $\{\hat{r}_i^{(0)}, \ldots, \hat{r}_i^{(K-1)}\}$ to its limit point and building a hyperplane passing through this limit point.

Alternatively, the authors in (Grothey *et al.* 1999, Grothey 2001) suggest a so-called *feasibility restoration* technique that tackles the worst-case scenario in a systematic manner. As the authors note, the failure of convergence in building V_i is due to the fact that a feasibility cut of the form in Eq. A.11 added in the master problem does not incur any new information about the effectiveness of the last change made in $\hat{r}_i$ towards optimality of the global solution. The key idea behind their feasibility restoration technique is to solve again the infeasible sub-problem with an artificially enlarged feasible region that includes the infeasible solution within its interior. The modification then recovers a value of $\alpha_i^{(K)}$ and $\lambda_i^{(K)}$ and provides the necessary cost objective information for algorithm to continue further. Thus, if

$$\hat{r}_i^{(k-1)^+} \equiv \max\left(0, r_i(x_i^{(k)}, u_i^{(k)}) - \hat{r}_i^{(k-1)}\right)$$

denotes the vector of infeasibility in the resource constraints for a given value of $\hat{r}_i^{(k-1)}$ (where maximum is taken componentwise), then the following sub-problem is feasible:

$$
\begin{aligned}
\alpha_i^{(K)^+} &\triangleq \underset{x_i, u_i}{\text{minimise}} \ f_i(x_i, u_i) && \text{(A.13)}\\
\text{s.t.} \quad & x_i = h_i(x_i, u_i)\\
& x_i \in X_i, u_i \in U_i\\
& r_i(x_i, u_i) \leq \hat{r}_i^{(K-1)} + m_1 \hat{r}_i^{(K-1)^+},
\end{aligned}
$$

where $m_1 > 0$ to ensure that the resulting sub-problem has at least one feasible solution strictly interior to its domain. The solution $\alpha_i^{(K)^+}$ and $\lambda_i^{(K)^+}$ of this modified sub-problem can then be used to build an additional optimality cut

$$\alpha_i^{(K)^+} - \left(\lambda_i^{(K)^+}\right)^T \left(\hat{r}_i - \left\{\hat{r}_i^{(K-1)} + m_1 \hat{r}_i^{(K-1)^+}\right\}\right)$$

in the master problem apart from the usual feasibility cut as in Eq. A.11. The remainder of the algorithm then proceeds according to Algorithm A.1 together with the scheme for building V_i as discussed here.

A.2.3 Relevant Research

The word *primal* in primal decomposition refers to the use of primal interaction variables (*e.g.*, $\hat{r}_i$) in coordinating the sub-problems. In an alternative *dual* coordination scheme, the dual variables, *i.e.*, the Lagrange multipliers, are used for the same coordination purpose. The first primal coordination scheme is due to Benders (1962), while the first dual scheme is due to Dantzig & Wolfe (1961). Kornai & Liptak (1965) applied Benders's (1962) algorithm for solving a large-scale linear program for Hungarian national planning bureau. The primal decomposition algorithm described here was taken from Geoffrion (1970). Geoffrion (1972) later extended much of the earlier work on the topic of primal decomposition in his so-called *Generalised Benders Decomposition* (GBD) scheme for solving large-scale non-linear programs. Geoffrion's work was pioneering in that it was later extended to various applications, for instance, in multicommodity distribution systems (Geoffrion & Graves 1974), process engineering (Takama & Umeda 1980), power systems (Alguacil & Conejo 2000), water resources management (Cai, McKinney, Lasdon & Watkins 2001) and communication networks (Mahey, Benchakroun & Boyer 2001). Molina (1979) surveyed the early research on both types of decomposition schemes. Various theoretical aspects on (generalised) primal decomposition were examined in (Bagajewicz & Manousiouthakis 1991, Sahinidis & Grossmann 1991, Grothey *et al.* 1999, Wu, Hartman & Wilson 2003). In process applications, Grossmann and his co-workers applied several variants of primal algorithm and its extension (in the form of so-called *generalised outer approximation*) for solving large-scale and/or non-linear problems of mixed-integer nature in process synthesis. See Grossmann & Daichendt (1996) for a review of related research.

B

Appendix to Chapter 6 – Implementation of Distributed Coordination Algorithm

This appendix describes an implementation of the distributed coordination algorithm (Algorithm 6.3) discussed in Chapter 6. The prototype was developed using MATLAB® software and its Optimisation Toolbox. MATLAB® was chosen as the preferred tool because of the different vector and matrix manipulation facilities that are available therein. The QP routine from Optimisation toolbox was used to solve all quadratic programs in Algorithm 6.3. Development of a separate QP solver that best fits with the requirements of coordination algorithm was not considered necessary as the MATLAB QP routine provides the sufficient information for our implementation.

B.1 Data Structures

The prototype was developed using the framework of Prob. 6.1. The unit elements were numbered using indexing method specified in Section 6.6. Different components involved in the problem formulation were modelled using appropriate data structures and cell arrays in MATLAB. A brief description of these data structures and cell arrays is given below.

B.1.1 MtrlType and UnitType

To model the P-graph form of the process, two different data structures namely – MtrlType and UnitType – were considered, MtrlType representing the material nodes and UnitType representing the unit elements. Tables B.1 and B.2 depict their formulation.

The sets of material and unit nodes in a P-graph were represented as MATLAB arrays mtrl or unit. Each element in these arrays is modelled using appropriate MtrlType or UnitType data structure.

Table B.1. MtrlType structure

MtrlType.id	% Material ID
MtrlType.UnitType_in	% UnitTypes which consume it
MtrlType.UnitType_out	% UnitTypes which produce it
MtrlType.rmtrlflag	% Raw-material flag, = 1 if raw-material, = 0 otherwise
MtrlType.prdctflag	% Product flag, = 1 if product, = 0 otherwise

Table B.2. UnitType structure

UnitType.UnitType_id	% UnitType ID
UnitType.type	% Type of unit = 'interim', 'source', 'sink'
UnitType.mtrl_in	% ID of incoming materials
UnitType.mtrl_out	% ID of outgoing materials
UnitType.unit_qty	% Total number of units of this type
UnitType.unit_id	% Id of actual units of this type
UnitType.CustCnxn	% Customer connection object
UnitType.SuppCnxn	% Supplier connection object
UnitType.Q	% Unit's cost objective coefficients
UnitType.c	%
UnitType.eqcon	% Local equality constraints
UnitType.ineqcon	% Local inequality constraints (not used at present)
UnitType.xss	% State variables
UnitType.uin	% Input flow-rates
UnitType.uutil	% Other local variables/degrees of freedom
UnitType.vout	% Demand disturbance at outlet

B.1.2 CustCnxn and SuppCnxn

The piping connections between unit elements were modelled using two data structures CustCnxn (defining the connection to a customer unit downstream) and SuppCnxn (defining the connection to a supplier unit upstream). Tables B.3 and B.4 depict the structure of both connection entities.

Each unit element may be connected to multiple supplier and customer elements. The associated instances of CustCnxn and SuppCnxn are combined into appropriate cell arrays and attached to corresponding instance of Unit-Type data structure (see Table B.2). The number of instances of CustCnxn or SuppCnxn correspond to the number of entries in output-set (mat_i^{out}) and input-set (mat_i^{in}) of materials of specific unit element.

B.1.3 Product Recipe and Mapping onto UnitType

The product recipes of product elements were modelled using a cell array recipe. Each element in recipe refers to an individual product and comprises a sequence of recipe tasks modelled as a combination of incoming and outgoing

Table B.3. CustCnxn structure

CustCnxn.UnitType % UnitType linked to this connection
CustCnxn.mtrl   % Type of material associated
CustCnxn.min_flow % Minimum flow
CustCnxn.max_flow % Maximum flow
CustCnxn.CustUnit % Details of customer UnitTypes and units
CustCnxn.eqcon   % Output dynamics

Table B.4. SuppCnxn structure

SuppCnxn.UnitType % UnitType linked to this connection
SuppCnxn.mtrl   % Type of material associated
SuppCnxn.min_flow % Minimum flow
SuppCnxn.max_flow % Maximum flow
SuppCnxn.SuppUnit % Details of supplier UnitTypes and units

materials. Two other variable arrays rmtrl and prdct were used to store the indices of incoming raw-materials and outgoing end-products.

B.2 Unit Module

Central to the implementation of Algorithm 6.3 was a *unit* module written as a MATLAB sub-routine. Table B.5 depicts the pseudo-code version of unit module. Note that this unit module is generic in that it can be applied to any unit element placed in any arbitrary network configuration.

B.3 Overall Implementation

Fig. B.1 outlines the overall implementation of the prototype. The procedure runs in two parts: (i) initial definition phase, where all data structures and cell arrays are defined, and (ii) execution of the algorithm.

To verify the optimality of the resulting solution, we also form an equivalent centralised problem and solve it using MATLAB QP routine. The aim is that the solution obtained from distributed algorithm must match that from the centralised algorithm. On occasions, we measure the number of flops (floating point operations) involved in both cases to make a comparison of the computational load.

Table B.5. Pseudo-code programme listing for Unit module

Shared memory variables: $\hat{v}_i$, z_i, α_i, $\alpha_{ij,updt}$, λ_i for all units i in the network. The variables are indexed by iteration count K in order to store new values at each subsequent iteration.

Local variables: x_i, u_i

for each CustCnxn **do**

    Retrieve Customer unit_ids

    Construct A_i, B_i and $A_{i,loc}, B_{i,loc}$ matrices from CustCnxn.eqcon

    Construct b_i vector by combining total demand from all customer units

end for

for $k = 1$ to K **do**

    for each SuppCnxn **do**

        for all SuppUnits indexed by j **do**

            Retrieve Supplier unit_ids

            Retrieve $\alpha_{ji}^{(k)}$, $\alpha_{ji,updt}^{(\mathbf{K})}$, $\lambda_{ji}^{(k)}$, $z_{ji}^{(k)}$, $k \in \mathbf{K}$

            Construct optimality cuts for iteration $k \in \mathbf{K}$ using this information

        end for

    end for

end for

Solve sub-problem $SP_i^{(K)}$ associated with the current unit i

for each supplier unit j **do**

    Update $\hat{v}_{ij}^{(K+1)}$ to be passed to unit j

end for

for all supplier units indexed by s **do**

    for $p = 1$ to K **do**

        Calculate total updates received from all suppliers $\alpha_{s,updt,supp}^{(p)} \triangleq \sum_s \alpha_{s,updt}^{(p)}$

    end for

end for

for each customer unit j **do**

    Update $z_{ij}^{(k)}$ and $\alpha_{ij}^{(k)}$ to be passed to j

    for $p = 1$ to K **do**

        Calculate total $\alpha_{ij,updt}^{(p)} \triangleq \alpha_{s,updt,supp}^{(p)} + \alpha_{ij,updt,loc}^{(p)}$

    end for

    Pass the whole vector $\alpha_{ij,updt}^{(\mathbf{K})}$, the vector of cut updates, to customer unit j

end for

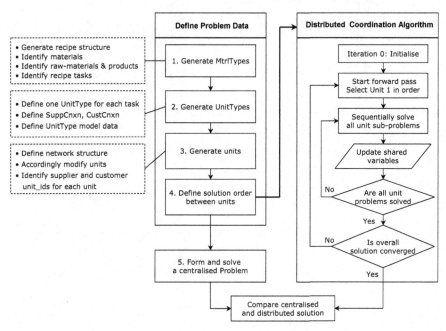

Fig. B.1. MATLAB implementation of distributed coordination algorithm

C

Appendix to Chapters 6 and 7 – Problem Data for Numerical Examples

This appendix provides the problem data for numerical examples discussed in Chapters 6 and 7. Tables C.1, C.3, C.5, C.7, C.9 and C.10 define this data for Examples 6.10, 6.11, 6.12, 6.13 in Chapter 6 and the multipurpose process example in Section 7.4.4 in Chapter 7.

Matrices $[B_i \ A_i]$ and $[B_{i,loc} \ A_{i,loc}]$ in the tables represent the dynamics equations for individual unit. The cost terms c_i refer to linear economic objectives such as the consumption of utilities (as a function of incoming flow-rates $u_{i,in}$ and states x_i) or the linear costs emerging from deviation terms $(u_i - \hat{u}_i)^2$ or $(x_i - \hat{x}_i)^2$.

We note the following points in reference to the data presented in this appendix:

- The problem data for all examples were generated using MATLAB's rand and randn functions. Care has been taken to ensure that resulting Q_i matrices in the objective functions $f_i(x_i, u_i)$ are strictly positive-definite, $i.e.$, the functions f_i are strictly convex.
- All units in the terminal tier ($i.e.$, those which produce the end-products) receive a demand $\hat{v}_j$ of 10 deviation units.
- Separate cost coefficients for raw-material supply are not used but it is assumed that such are included as part of the objective functions.
- The distributed algorithm is initialised using a feasible solution derived in a centralised manner by solving a linear program (using MATLAB's LP subroutine) with zero objective terms and a composite model of all process units. In principle, this problem can equally be solved in a distributed manner using the Algorithm 6.3.

Table C.1. Problem data for MIXER example (Example 6.10)

Variable Name	unit (1,1)	unit (2,1)	unit (1,2)
Q_i	$\begin{pmatrix} 0.8 & 0 & 0 \\ 0 & 0.9 & 0 \\ 0 & 0 & 0.5 \end{pmatrix}$	$\begin{pmatrix} 0.5 & 0 & 0 \\ 0 & 0.3 & 0 \\ 0 & 0 & 0.1 \end{pmatrix}$	$\begin{pmatrix} 0.3 & 0 & 0 & 0 \\ 0 & 0.2 & 0 & 0 \\ 0 & 0 & 1 & 0 \\ 0 & 0 & 0 & 0.2 \end{pmatrix}$
c_i^T	$(0.2\ 0.4\ 1.5)$	$(-0.3\ 0.7\ 0.8)$	$(1\ -0.9\ 0.3\ 0.3)$
$[B_i \quad A_i]$	$(14.1\ 11.7\ 15.0)$	$(15.8\ 17.6\ 14.5)$	$(0.5\ 0.5\ 16.9\ 3.8)$
$[B_{i,loc} \quad A_{i,loc}]$	$(18.9\ 17.7\ 7.6)$	$(0.4\ 7.1\ 19.4)$	$(3.2\ 3.2\ 17.6\ 19.9)$

Table C.2. Progress of iterations in MIXER example (Example 6.10)

Iteration	1	2	3	4	5	6	Centralised
$u_{(1,1,1)}$	0.3250	0.3236	0.3237	0.3237	0.3237	0.3237	0.3237
$x_{(1,1,1)}$	-0.1631	-0.1650	-0.1649	-0.1649	-0.1649	-0.1649	-0.1649
$x_{(1,1,2)}$	-0.4284	-0.4206	-0.4208	-0.4208	-0.4208	-0.4208	-0.4208
$u_{(2,1,1)}$	0.9793	0.9785	0.9785	0.9785	0.9785	0.9785	0.9785
$x_{(2,1,1)}$	-0.8651	-0.8661	-0.8661	-0.8661	-0.8661	-0.8661	-0.8661
$x_{(2,1,2)}$	0.2964	0.2968	0.2968	0.2968	0.2968	0.2968	0.2968
$u_{(1,2,1)}$	-3.6762	-3.6777	-3.6777	-3.6777	-3.6777	-3.6777	-3.6777
$u_{(1,2,2)}$	4.5199	4.5201	4.5201	4.5201	4.5201	4.5201	4.5201
$x_{(1,2,1)}$	0.7455	0.7455	0.7455	0.7455	0.7455	0.7455	0.7455
$x_{(1,2,2)}$	-0.7950	-0.7948	-0.7948	-0.7948	-0.7948	-0.7948	-0.7948
Total Cost	-4.2441	-4.1877	-4.1877	-4.1877	-4.1877	-4.1877	-4.1877

Table C.3. Problem data for SPLITTER example (Example 6.11)

Variable Name	unit (1,1)	unit (1,2)	unit (2,2)
Q_i	$\begin{pmatrix} 3.6 & 0 & 0 \\ 0 & 2.3 & 0 \\ 0 & 0 & 3.1 \end{pmatrix}$	$\begin{pmatrix} 2.3 & 0 & 0 \\ 0 & 1.5 & 0 \\ 0 & 0 & 2.1 \end{pmatrix}$	$\begin{pmatrix} 0.23 & 0 & 0 \\ 0 & 0.15 & 0 \\ 0 & 0 & 0.21 \end{pmatrix}$
c_i^T	$(-1.1\ -0.6\ -0.2)$	$(0.7\ -1.2\ -0.5)$	$(-0.4\ 1.0\ 1.8)$
$[B_i \quad A_i]$	$(6.3\ 1.1\ 2.0)$	$(2.7\ 2.6\ 4.0)$	$(4.0\ -4.2\ -1.5)$
$[B_{i,loc} \quad A_{i,loc}]$	$(-6.8\ 0.9\ -1.7)$	$(0.6\ 2.9\ -0.1)$	$(0.7\ -0.4\ -5.1)$

Table C.4. Progress of iterations in SPLITTER example (Example 6.11)

Iteration	1	2	3	4	7	10	12	Centralised
$u_{(1,1,1)}$	0.0191	0.0523	0.0290	0.0338	0.0304	0.0306	0.0307	0.0307
$x_{(1,1,1)}$	0.0150	0.4065	0.1314	0.1890	0.1485	0.1510	0.1516	0.1515
$x_{(1,1,2)}$	-0.0684	0.0062	-0.0462	-0.0353	-0.0430	-0.0425	-0.0424	-0.0424
$u_{(1,2,1)}$	0.6317	0.4718	0.5841	0.5606	0.5771	0.5761	0.5759	0.5759
$x_{(1,2,1)}$	-0.0579	-0.0219	-0.0472	-0.0419	-0.0456	-0.0454	-0.0453	-0.0453
$x_{(1,2,2)}$	2.1112	2.1958	2.1364	2.1488	2.1401	2.1406	2.1407	2.1407
$u_{(2,2,1)}$	0.1571	-0.2373	-0.2337	-0.2917	-0.2911	-0.2996	-0.3010	-0.3010
$x_{(2,2,1)}$	-2.3036	-2.6701	-2.6667	-2.7206	-2.7201	-2.7280	-2.7293	-2.7293
$x_{(2,2,2)}$	0.2022	0.1769	0.1771	0.1734	0.1734	0.1728	0.1727	0.1727
Total Cost	2.7059	2.6980	2.7676	2.7694	2.7710	2.7710	2.7710	2.7710

Table C.5. Problem data for SPLITTER-STAIRCASE example (Example 6.12)

Variable Name	unit (1,3)	unit (2,3)
Q_i	$\begin{pmatrix} 2.2 & 0 & 0 \\ 0 & 3.1 & 0 \\ 0 & 0 & 2.8 \end{pmatrix}$	$\begin{pmatrix} 1.7 & 0 & 0 \\ 0 & 1.3 & 0 \\ 0 & 0 & 1.9 \end{pmatrix}$
c_i^T	$(-2.4\ -0.2\ -0.1)$	$(-0.4\ -0.2\ -1.5)$
$[B_i \quad A_i]$	$(2.0\ -1.9\ -1.1)$	$(0.2\ -7.5\ -3.0)$
$[B_{i,loc} \quad A_{i,loc}]$	$(-1.1\ 0.8\ -1.5)$	$(3.8\ 1.8\ 2.9)$

The problem data for unit elements $(1, 1)$, $(1, 2)$ and $(2, 2)$ are taken from those in Table C.3 for Example 6.11.

Table C.6. Progress of iterations in SPLITTER-STAIRCASE example (Example 6.12)

Iteration	1	2	3	4	6	8	10	Centralised
$u_{(1,1,1)}$	0.0191	-0.0099	0.0064	0.0090	0.0076	0.0074	0.0073	0.0073
$x_{(1,1,1)}$	0.0150	-0.3268	-0.1345	-0.1037	-0.1207	-0.1234	-0.1238	-0.1239
$x_{(1,1,2)}$	-0.0684	-0.1335	-0.0969	-0.0910	-0.0943	-0.0948	-0.0949	-0.0949
$u_{(1,2,1)}$	0.0357	0.0788	0.0024	-0.0106	-0.0037	-0.0026	-0.0024	-0.0024
$x_{(1,2,1)}$	0.0220	0.0035	0.0209	0.0238	0.0222	0.0220	0.0219	0.0219
$x_{(1,2,2)}$	0.8523	0.5744	0.6205	0.6264	0.6227	0.6221	0.6220	0.6220
$u_{(2,2,1)}$	-0.7245	-0.3801	-0.2417	-0.2727	-0.2767	-0.2774	-0.2775	-0.2775
$x_{(2,2,1)}$	-0.8261	-0.4552	-0.3269	-0.3559	-0.3596	-0.3602	-0.3603	-0.3603
$x_{(2,2,2)}$	-0.0346	-0.0165	-0.0075	-0.0095	-0.0098	-0.0098	-0.0098	-0.0098
$u_{(1,3,1)}$	2.5196	2.5426	2.5388	2.5383	2.5386	2.5387	2.5387	2.5387
$x_{(1,3,1)}$	-1.1776	-1.1516	-1.1559	-1.1565	-1.1562	-1.1561	-1.1561	-1.1561
$x_{(1,3,2)}$	-2.4758	-2.4788	-2.4783	-2.4782	-2.4783	-2.4783	-2.4783	-2.4783
$u_{(2,3,1)}$	0.4161	0.4177	0.4183	0.4181	0.4181	0.4181	0.4181	0.4181
$x_{(2,3,1)}$	-1.4688	-1.4676	-1.4672	-1.4673	-1.4673	-1.4673	-1.4673	-1.4673
$x_{(2,3,2)}$	0.3665	0.3636	0.3626	0.3629	0.3629	0.3629	0.3629	0.3629
Total Cost	12.7200	12.8080	12.8409	12.8409	12.8413	12.8413	12.8413	12.8413

Table C.7. Problem data for MULTIPROD-SPLITTER example (Example 6.13)

Variable Name	unit (1,1)
Q_i	$\begin{pmatrix} 3.6 & 0 & 0 & 0 \\ 0 & 2.3 & 0 & 0 \\ 0 & 0 & 3.1 & 0 \\ 0 & 0 & 0 & 3.5 \end{pmatrix}$
c_i^T	$(-1.1\ -0.6\ -0.2\ 1.5)$
$[B_i \quad A_i]$	$(6.3\ 1.1\ 2.0\ -3.5)$
$[B_{i,loc} \quad A_{i,loc}]$	$(-6.8\ 0.9\ -1.7\ 1.8)$

The problem data for unit elements $(1,2)$ and $(2,2)$ are taken from those in Table C.3 for Example 6.11 while that for unit element $(3,2)$ from Table C.5 for element $(2,3)$.

Table C.8. Progress of iterations in MULTIPROD-SPLITTER example (Example 6.13)

Iteration	1	2	3	4	7	10	13	Centralised
$u_{(1,1,1)}$	0.0285	-0.0266	0.0008	-0.0077	-0.0030	-0.0031	-0.0031	-0.0031
$x_{(1,1,1)}$	-0.0078	0.3239	0.0987	0.1678	0.1289	0.1303	0.1300	0.1300
$x_{(1,1,2)}$	-0.1680	-0.0535	-0.1118	-0.0939	-0.1039	-0.1035	-0.1036	-0.1036
$x_{(1,1,3)}$	-0.0472	-0.3131	-0.1521	-0.2016	-0.1738	-0.1748	-0.1746	-0.1746
$u_{(1,2,1)}$	0.6738	0.5185	0.6170	0.5868	0.6038	0.6032	0.6033	0.6033
$x_{(1,2,1)}$	-0.0674	-0.0324	-0.0546	-0.0478	-0.0516	-0.0515	-0.0515	-0.0515
$x_{(1,2,2)}$	2.0890	2.1711	2.1190	2.1350	2.1260	2.1263	2.1262	2.1262
$u_{(2,2,1)}$	0.5036	-0.0963	0.0369	-0.0791	-0.0721	-0.0760	-0.0760	-0.0760
$x_{(2,2,1)}$	-1.9815	-2.5391	-2.4153	-2.5230	-2.5165	-2.5202	-2.5202	-2.5202
$x_{(2,2,2)}$	0.2245	0.1859	0.1945	0.1870	0.1875	0.1872	0.1872	0.1872
$u_{(3,2,1)}$	0.3878	0.3930	0.3910	0.3916	0.3913	0.3913	0.3913	0.3913
$x_{(3,2,1)}$	-1.4896	-1.4857	-1.4872	-1.4868	-1.4870	-1.4870	-1.4870	-1.4870
$x_{(3,2,2)}$	0.4165	0.4072	0.4108	0.4097	0.4103	0.4103	0.4103	0.4103
Total Cost	3.8165	3.8455	3.9382	3.9393	3.9428	3.9429	3.9429	3.9429

Table C.9. Problem data for multipurpose process example (Section 7.4.4)

Unit	diag(Q_i)	c_i^T
1	1.9 2.5 2.3 2.0 2.3	-4.2 -0.3 0.1 -0.5 0.2
2	2.5 2.2 3.1 2.6 2.9 2.9 2.7	-2.7 -0.8 -3.3 -2.2 0.5 0 -0.7
3	2.7 2.4 2.9 2.1 2.5	1.4 0.9 0 1.3 -2.7
4	2.3 2.7 3.0 2.5 2.6	2.5 0.7 2.3 -2.8 1
5	1.8 3.0 2.0	2.6 -1.9 -2.4
6	3.5 3.9 2.0	-3 -1.5 1.5
7	3.0 2.7 2.3	0.2 -4 -3.5
8	1.7 1.7 1.9 1.9	1 1 -1.8 -0.7
9	1.8 2.4 2.4 2.2 2.9	0.6 1.2 -0.6 3.6 4.1
10	2.3 2.1 2.5 2.2 2.0	-5.1 -0.7 -6.7 1.3 0
11	1.8 1.6 1.8 1.9 1.9	-2.1 -0.2 0.3 -3.2 2.1
12	1.7 1.7 1.6 1.6	-2.2 -2.2 0.6 -1.4
13	1.7 1.7 1.6 1.7	-2.2 -2.2 0.9 -1.1
14	1.8 1.5 2.5	-2.1 1.1 -1.2
15	1.7 1.7 1.8 1.8	-0.3 -0.3 1.8 0.1
16	2.5 2.2 2.5	0.5 1.1 3.9
17	1.7 1.7 1.7 1.6	1.6 1.6 -0.3 2.4

Table C.10. Problem data for multipurpose process case example (Section 7.4.4)

Unit	$[B_i \quad A_i]$	$[B_{i,loc} \quad A_{i,loc}]$
1	6.4 0.6 0.9 4.5 2.6	-3 7.5 -0.1 1 0.7
		-10.3 0.3 0.9 3.8 1.5
2	-4.9 -4.4 2.1 -2.5 2.1 1.5 3.9	-4.7 4.1 -6.5 1.2 6 4.2 6.6
		0.5 0.6 2 -2.9 3.9 0.5 6.2
		1.8 13.7 2.5 0.2 -13.2 4.3 4.8
3	-8.2 -4.8 0.2 2.4 1.5	-4.9 -3 5.4 -8.1 -0.7
		3.5 3.5 -6.7 -7.2 -5.6
4	-1.4 1.3 -10.1 3.8 -5.3	-2.9 -7.4 2.7 -3.4 4.5
		-4.4 1.6 1.8 3.5 -10.6
5	1.5 -3.8 -1.6	-3.6 0.8 4.9
6	-0.5 -2.3 2	5.1 0.4 5.6
7	3.2 -6.8 1.7	-1.4 -3.4 -4.9
8	1.5 1.5 1.3 2.1	5.6 5.6 0.9 6.1
9	7.3 3 4.3 3.6 7.7	-5.1 -1.3 -4.2 3.8 -8
		1.1 12.5 4.1 -8.5 5.6
10	-5.5 4.1 -0.5 -4.5 -2	2 0.2 -4 3 0.4
		4.9 -4.6 -8.3 2.8 2.3
11	1 -11.3 3.4 0.9 6.7	1.3 1.7 -2.9 4.3 -0.3
		10.5 1.5 4.5 4.9 6.6
12	1.2 1.2 -0.7 -8.6	-7.1 -7.1 -2.5 -2
13	-3 -3 1.7 1.5	3.7 3.7 4.5 -0.7
14	-0.4 5.9 -6.7	1.5 4.1 0.7
15	-1.1 -1.1 -1.5 -0.8	2.9 2.9 5.7 -7.3
16	7 2.9 4.2	2.3 -3.4 -11.1
17	5.5 5.5 -8 1.1	1.5 2.3 -6.1 1.2

Table C.11. Progress of iterations for u_i variables in multipurpose process example (Section 7.4.4)

Iteration	1	2	3	5	8	12	15	Centralised
$u_{1,1}$	0.0095	0.0456	0.0436	0.0433	0.0432	0.0431	0.0431	0.0431
$u_{1,2}$	0.0076	0.0082	0.0081	0.0081	0.0081	0.0081	0.0081	0.0081
$u_{2,1}$	0.1328	0.0941	0.0956	0.0957	0.0957	0.0957	0.0957	0.0957
$u_{2,2}$	-0.1431	-0.1659	-0.1650	-0.1650	-0.1650	-0.1650	-0.1650	-0.1650
$u_{2,3}$	0.5550	0.5670	0.5665	0.5664	0.5664	0.5664	0.5664	0.5664
$u_{3,1}$	0.0411	0.0074	0.0293	0.0269	0.0270	0.0271	0.0271	0.0271
$u_{3,2}$	0.0308	0.0142	0.0250	0.0238	0.0239	0.0239	0.0239	0.0239
$u_{4,1}$	-0.3654	-0.3524	-0.3525	-0.3522	-0.3522	-0.3522	-0.3522	-0.3522
$u_{4,2}$	0.1104	0.1277	0.1276	0.1280	0.1280	0.1280	0.1280	0.1280
$u_{5,1}$	-0.1527	-0.1622	-0.1628	-0.1624	-0.1622	-0.1622	-0.1622	-0.1622
$u_{6,1}$	0.3582	0.1253	0.1477	0.1486	0.1491	0.1492	0.1492	0.1492
$u_{7,1}$	-0.0825	-0.0349	-0.0398	-0.0389	-0.0389	-0.0389	-0.0389	-0.0389
$u_{8,1}$	-0.1921	-0.4540	-0.2680	-0.3163	-0.3300	-0.3306	-0.3306	-0.3306
$u_{8,2}$	-0.2702	-0.5528	-0.3971	-0.4346	-0.4428	-0.4431	-0.4431	-0.4431
$u_{9,1}$	0.8506	0.8216	0.8083	0.8122	0.8133	0.8133	0.8134	0.8134
$u_{9,2}$	-0.0292	-0.0445	-0.0641	-0.0591	-0.0576	-0.0576	-0.0576	-0.0576
$u_{10,1}$	1.2992	1.1814	1.1834	1.1848	1.1851	1.1851	1.1851	1.1851
$u_{10,2}$	0.7731	0.6854	0.6998	0.6996	0.6998	0.6998	0.6998	0.6998
$u_{11,1}$	0.3484	0.3348	0.3337	0.3337	0.3337	0.3337	0.3337	0.3337
$u_{11,2}$	-0.4351	-0.4772	-0.4863	-0.4863	-0.4863	-0.4863	-0.4863	-0.4863
$u_{12,1}$	0.1609	0.1461	0.1624	0.1566	0.1538	0.1536	0.1536	0.1536
$u_{12,2}$	0.0828	0.0474	0.0333	0.0384	0.0409	0.0411	0.0411	0.0411
$u_{13,1}$	-0.0240	0.1611	0.1514	0.1517	0.1517	0.1517	0.1517	0.1517
$u_{13,2}$	0.4433	0.5912	0.5883	0.5883	0.5884	0.5884	0.5884	0.5884
$u_{14,1}$	1.4841	0.5846	1.0408	0.9063	0.8679	0.8664	0.8664	0.8664
$u_{15,1}$	-1.4539	-1.4619	-1.4606	-1.4608	-1.4609	-1.4609	-1.4609	-1.4609
$u_{15,2}$	-1.6569	-1.5965	-1.5985	-1.5984	-1.5984	-1.5984	-1.5984	-1.5984
$u_{16,1}$	1.1624	1.1627	1.1627	1.1627	1.1627	1.1627	1.1627	1.1627
$u_{17,1}$	0.0349	0.0270	0.0249	0.0250	0.0250	0.0250	0.0250	0.0250
$u_{17,2}$	0.1560	0.1252	0.1282	0.1281	0.1281	0.1281	0.1281	0.1281
Total Cost	3.2155	4.2393	4.6262	4.7294	4.7339	4.7340	4.7340	4.7340

Table C.12. Progress of iterations for x_i variables in multipurpose process example (Section 7.4.4)

Iteration	1	2	3	5	8	12	15	Centralised
$x_{1,1}$	-0.0463	-0.0315	-0.0323	-0.0325	-0.0325	-0.0325	-0.0325	-0.0325
$x_{1,2}$	0.1257	0.2054	0.2010	0.2003	0.2000	0.2000	0.2000	0.2000
$x_{1,3}$	-0.2268	-0.1902	-0.1922	-0.1925	-0.1926	-0.1926	-0.1926	-0.1926
$x_{2,1}$	1.0867	1.0790	1.0793	1.0793	1.0793	1.0793	1.0793	1.0793
$x_{2,2}$	0.1587	0.1395	0.1402	0.1403	0.1403	0.1403	0.1403	0.1403
$x_{2,3}$	0.2800	0.2932	0.2927	0.2926	0.2926	0.2926	0.2926	0.2926
$x_{2,4}$	0.2100	0.2189	0.2185	0.2185	0.2185	0.2185	0.2185	0.2185
$x_{3,1}$	-0.3697	-0.3982	-0.3797	-0.3817	-0.3816	-0.3816	-0.3816	-0.3816
$x_{3,2}$	-0.3655	-0.3572	-0.3626	-0.3620	-0.3620	-0.3620	-0.3620	-0.3620
$x_{3,3}$	0.9572	0.9492	0.9544	0.9538	0.9538	0.9538	0.9538	0.9538
$x_{4,1}$	0.1144	0.1618	0.1613	0.1625	0.1625	0.1625	0.1625	0.1625
$x_{4,2}$	0.7293	0.7219	0.7220	0.7218	0.7218	0.7218	0.7218	0.7218
$x_{4,3}$	0.4286	0.4314	0.4313	0.4314	0.4314	0.4314	0.4314	0.4314
$x_{5,1}$	-0.0140	-0.0061	0.0721	0.0473	0.0341	0.0351	0.0351	0.0351
$x_{5,2}$	-0.1099	-0.1182	-0.1314	-0.1270	-0.1247	-0.1249	-0.1249	-0.1249
$x_{6,1}$	-0.3118	-0.6389	-0.6057	-0.6041	-0.6032	-0.6033	-0.6033	-0.6033
$x_{6,2}$	-0.3040	-0.0684	-0.0912	-0.0922	-0.0928	-0.0928	-0.0927	-0.0927
$x_{7,1}$	-0.0268	-0.1088	-0.0994	-0.1010	-0.1009	-0.1009	-0.1009	-0.1009
$x_{7,2}$	0.0422	0.0854	0.0804	0.0812	0.0812	0.0812	0.0812	0.0812
$x_{8,1}$	-0.1998	1.0638	0.3030	0.4924	0.5469	0.5409	0.5406	0.5406
$x_{8,2}$	0.4539	0.7673	0.5659	0.6167	0.6322	0.6306	0.6305	0.6305
$x_{9,1}$	0.6172	0.6404	0.6346	0.6353	0.6355	0.6355	0.6355	0.6355
$x_{9,2}$	-0.2951	-0.3085	-0.3430	-0.3346	-0.3320	-0.3321	-0.3321	-0.3321
$x_{9,3}$	-1.0017	-0.9993	-1.0010	-1.0006	-1.0005	-1.0005	-1.0005	-1.0005
$x_{10,1}$	-0.0032	0.0276	0.0109	0.0120	0.0120	0.0120	0.0120	0.0120
$x_{10,2}$	-0.9043	-0.7842	-0.8079	-0.8073	-0.8076	-0.8076	-0.8076	-0.8076
$x_{10,3}$	-0.1324	-0.0917	-0.0987	-0.0987	-0.0988	-0.0988	-0.0988	-0.0988
$x_{11,1}$	0.5779	0.5732	0.5716	0.5716	0.5716	0.5716	0.5716	0.5716
$x_{11,2}$	0.3776	0.3966	0.3997	0.3997	0.3997	0.3997	0.3997	0.3997
$x_{11,3}$	-1.1298	-1.1095	-1.1068	-1.1067	-1.1068	-1.1068	-1.1068	-1.1068
$x_{12,1}$	-0.7694	-0.7556	-0.7633	-0.7611	-0.7599	-0.7600	-0.7600	-0.7600
$x_{12,2}$	0.0966	0.2576	0.2594	0.2590	0.2589	0.2589	0.2589	0.2589
$x_{13,1}$	-0.1822	-0.4730	-0.4621	-0.4625	-0.4625	-0.4625	-0.4625	-0.4625
$x_{13,2}$	1.0451	0.9361	0.9387	0.9386	0.9386	0.9386	0.9386	0.9386
$x_{14,1}$	-0.2373	0.0408	-0.1003	-0.0587	-0.0457	-0.0463	-0.0463	-0.0463
$x_{14,2}$	-1.7901	-1.4915	-1.6430	-1.5983	-1.5843	-1.5850	-1.5851	-1.5851
$x_{15,1}$	-2.6308	-2.6657	-2.6652	-2.6652	-2.6651	-2.6651	-2.6651	-2.6651
$x_{15,2}$	-3.2900	-3.2964	-3.2964	-3.2963	-3.2963	-3.2963	-3.2963	-3.2963
$x_{16,1}$	0.5278	0.5261	0.5264	0.5264	0.5264	0.5264	0.5264	0.5264
$x_{16,2}$	0.0792	0.0798	0.0797	0.0797	0.0797	0.0797	0.0797	0.0797
$x_{17,1}$	-0.2482	-0.2600	-0.2596	-0.2596	-0.2596	-0.2596	-0.2596	-0.2596
$x_{17,2}$	-1.6041	-1.5954	-1.5963	-1.5963	-1.5963	-1.5963	-1.5963	-1.5963

References

Aldea, A., Bañares Alcaántara, R., Jiménez, L., Moreno, A., Martínez, J. and Riaño, D. (2004). The scope of application of multi-agent systems in the process industry: Three case studies, *Expert Systems with Applications* **26**: 39–47.

Alguacil, N. and Conejo, A. (2000). Multiperiod optimal power flow using benders decomposition, *IEEE Transactions on Power Systems* **15**: 196–201.

Alkaya, D., Vasantharajan, S. and Biegler, L. (2000). Generalization of a tailored approach for process optimization, *Industrial and Engineering Chemistry Research* **39**: 1731–1742.

Altman, E., Başar, T. and Srikant, R. (2002). Nash equilibria for combined flow control and routing in networks: Asymptotic behavior for a large number of users, *IEEE Transactions on Automatic Control* **47**(6): 917–930.

Amdahl, G., Blaauw, G. and Brooks, F. J. (1964). Architecture of the ibm system/360, *IBM Journal of Research and Development* **8**: 87–101. Reprinted in Vol. 44 No. 1/2, Jan/Mar 2000.

Anderson, J. (1997). Future directions of R & D in the process industries, *Computers in Industry* **34**: 161–172.

Androulakis, I. and Reklaitis, G. (1999). Approaches to asynchronous decentralized decision making, *Computers and Chemical Engineering* **23**: 341–355.

ANSI/ISA (1995). Batch control, Part 1: Models and terminilogy, *International Standard ANSI/ISA-S88.01*, The Instrumentation Systems and Automation Society, North Carolina, USA.

Askin, R., Ciarallo, F. and Lundgren, N. (1999). An empirical evaluation of holonic and fractal layouts, *International Journal of Production Research* **37**(5): 961–978.

Babiceanu, R. and Chen, F. (2006). Development and applications of holonic manufacturing: A survey, *Journal of Intelligent Manufacturing* **17**: 111–131.

Backx, T., Bosgra, O. and Marquardt, W. (2000). Integration of model predictive control and optimization of processes, *Proceedings of ADCHEM*.

Bagajewicz, M. and Manousiouthakis, V. (1991). On the generalized benders decomposition, *Computers and Chemical Engineering* **15**(10): 691–700.

Basar, T. and Olsder, G. (1995). *Dynamic Non-cooperative Game Theory*, Academic Press, London.

Batres, R., Asprey, S., Fuchino, T. and Naka, Y. (1999). A KQML multi-agent environment for concurrent process engineering, *Computers and Chemical Engineering* **23, Supplement**: S653–S656.

Bazaraa, M., Sherali, H. and Shetty, C. (1993). *Nonlinear Programming, Theory and Algorithms*, Wiley-Interscience Seires in Discrete Mathematics and Optimization, 2nd edn, John Wiley and Sons, Inc., New York.

Beard, R., Lawton, J. and Hadaegh, F. (2001). A coordination architecture for spacecraft formation control, *IEEE Transactions on Control Systems Technology* **9**(6): 777–789.

Benders, J. (1962). Partioning procedures for solving mixed-variables programming problems, *Numerische Mathematik* **4**: 238–252.

Bertsekas, D. and Tsitsiklis, J. (1989). *Parallel and Distributed Computation: Numerical Methods*, Prentice Hall, New Jersey, USA.

Biegler, L., Grossmann, I. and Westerberg, A. (1997). *Systematic Methods of Chemical Process Design*, Physical and Chemical Engineering Series, Prentice Hall International Series, NJ, USA.

Bolton, L. and Perris, T. (1999). A vision of future industrial needs and capabilities: Process modelling, simulation and control, *Technical report*, CAPRI Project, http://cape-21.ucl.org.uk.

Bongaerts, L. (1998). *Integration of Scheduling and Control in Holonic Manufacturing Systems*, Phd thesis, Department Werktuigkunde Afdeling Productietechnieken, Katholieke Universiteit, Leuven, Belgium.

Bongaerts, L., Monostori, L., McFarlane, D. and Kádár (2000). Hierarchy in distributed shop floor control, *Computers in Industry* **43**: 123–137.

Booch, G. and Rumbaugh, J. (2005). *The Unified Modeling Language User Guide (OMG Technology)*, 2nd edn, Addison Wesley, Boston, USA.

Brennan, R., Fletcher, M. and Norrie, D. (2002). An agent-based approach to reconfiguration of real-time distributed control systems, *IEEE Transactions on Robotics and Automation* **18**(4): 444–451.

Brosilow, C. and Lasdon, L. (1965). A two level optimization technique for recycle processes, *AIChE Symposium Series* **4**: 75–83.

Bussmann, S. (2003). *An Agent-Oriented Design Methodology for Production Control*, Phd, Department of Electronics and Computer Science, University of Southampton, UK.

Bussmann, S., Jennings, N. and Wooldridge, M. (2001). *Agent-Oriented Software Engineering*, Vol. 1957 of *Lecture Notes in Computer Science*, Springer-Verlag, Berlin, chapter On the Identification of Agents in the Design of Production Control Systems, pp. 141–162.

Cai, X., McKinney, D., Lasdon, L. and Watkins, D. (2001). Solving large nonconvex water resources management models using generalized benders decomposition, *Operations Research* **49**(2): 235–245.

Camarinha-Matos, L., Afsarmanesh, H. and Rabelo, R. (2003). Infrastructure developments for agile virtual enterprises, *International Journal of Computer Integrated Manufacturing* **16**(4-5): 235–254.

Camponogara, E., Jia, D., Krogh, B. and Talukdar, S. (2002). Distributed model predictive control, *IEEE Control Systems Magazine* pp. 44–52.

Cefic (2006). Facts and figures: The european chemical industry in a worldwide perspective, *Technical report*, Cefic - European Chemical Industry Council, Brussels, Belgium.

Chaib-draa, B. and Müller, J. e. (2006). *Multiagent-based Supply Chain Management*, Vol. 28 of *Studies in Computational Intelligence*, Springer, Germany.

Chandler, A. (2005). *Shaping the Industrial Century: The Remarkable Story of the Evolution of the Modern Chemical and Pharmaceutical Industries*, Vol. 46 of *Harvard Studies in Business History*, Harvard University Press, MA, USA.

Cheung, H. M. E., Yeung, W. H. R., Ng, H. C. A. and Fung, S. T. R. (2000). HSCF: A holonic shopfloor control framework for flexible manufacturing systems, *International Journal of Computer Integrated Manufacturing* **13**(2): 121–138.

Chirn, J. and McFarlane, D. (2001). Holonic systems in todays factories: A migration strategy, *Journal of Applied Systems Studies* **2**(1): 82–105.

Chokshi, N., Matson, J, B. and McFarlane, D. (2000). Distributed co-ordination of steel-making operations for reduced production stoppages, *Proceedings of Manufacturing and Production Logistics and Management Conference*, Grenoble, France.

Chokshi, N. and McFarlane, D. (2002). *Multi-Agent-Systems and Applications II*, Mařík and Štěpáanková, O. and Krautwurmová, H. and Luck, M. (Eds.), Vol. 2322 of *Lecture Notes in Artificial Intelligence*, Springer-Verlag, Berlin, chapter Rationales for Holonic Applications in Chemical Process Industries, pp. 336–350.

Christensen, J. (1994). Holonic manufacturing systems - initial architecture and standards directions, *First European Conference on Holonic Manufacturing Systems, Germany* .

Dantzig, G. and Wolfe, P. (1961). The decomposition algorithm for linear programming, *Econometrica* **29**(4): 767–778.

Deen, S. (1993). Co-operation issues in holonic manufacturing systems, *in* H. Yoshikawa and Goossenaerts (eds), *Proc. of DIISM'93 - Design of Information Infrastructure Systems for Manufacturing*, Vol. B-14 of *IFIP Transactions*, Elsevier Science, Amsterdam, pp. 401–412.

Deonckheere, J., Disney, S., Lambrecht, M. and Towill, D. (2002). Transfer function analysis of forecasting induced bullwhip in supply chains, *International Journal of Production Economics* **78**: 133–144.

Dilts, D., Boyd, N. and Whorms, H. (1991). The evolution of control architectures for automated manufacturing systems, *Journal of Manufacturing Systems* **10**(1): 79–93.

Douglas, J. M. (1988). *Conceptual Design of Chemical Processes*, Mc-Graw Hill Chemical Engineering Series, Mc-Graw Hill Company, New York.

Duffie, N. and Piper, R. (1987). Non-hierarchical control of a flexible manufacturing cell, *Robotics and Computer Integrated Manufacturing* **3**(2): 175–179.

Duffie, N. and Prabhu, V. (1994). Real-time distributed scheduling of heterarchical manufacturing, *Journal of Manufacturing Systems* **13**.

Durfee, E. and Lesser, V. (1991). Partial global planning: A coordinatin framework for distributed hypothesis formation, *IEEE Transactions on Systems, Man, and Cybernetics* **21**(5): 1167–1183.

Edgar, T. (2004). Control and operations: When does controllability equal profitability?, *Computers and Chemical Engineering* **29**: 41–49.

Edgar, T. H., Himmelblau, D. and Lasdon, L. (2001). *Chemical Process Optimization*, Chemical Engineering Series, 2nd edn, McGraw-Hill Publications, New York, USA.

Egerstedt, M. and Hu, X. (2001). Formation constrained multi-agent control, *IEEE Transactions on Robotics and Automation* **17**(6): 947–951.

Eo, S., Chang, T., Shin, D. and Yoon, E. (2000). Cooperative problem solving in diagnostic agents for chemical processes, *Computers and Chemical Engineering* **24**: 729–734.

Felcht, U. (2002). The future shape of the process industries, *Chemical Engineering Technology* **25**(4): 345–355.

Fiacco, A. (1983). *Introduction to Sensitivity and Stability Analysis in Nonlinear Programming*, Vol. 165 of *Mathematics in Science and Engineering*, Academic Press, New York.

Fiacco, A. and Kyparisis, J. (1986). Convexity and concavity properties of the optimal value function in paramteric nonlinear programming, *Journal of Optimization Theory and Applications* **48**(1): 95–126.

Findeisen, W., Bailey, F., Brdys, M., Malinowski, K., Tatjewski, P. and Wozniak, A. (1980). *Control and Coordination in Hierarchical Systems*, Vol. 9 of *International Series on Applied Systems Analysis*, John Wiley And Sons, Chichester.

Fischer, K. (1999). Agent-based design of holonic manufacturing systems, *Robotics and Autonomous Systems* **27**: 3–13.

Fox, M., Barbuceanu, M. and Teigen, R. (2000). Agent-oriented supply-chain management, *The International Journal of Flexible Manufacturing Systems* **12**: 165–188.

Friedler, F., Tarján, K., Huang, Y. and Fan, L. (1992). Graph-theoretic approach to process synthesis: Axioms and theorems, *Chemical Engineering Science* **47**(8): 1973–1988.

García-Flores, R., Wang, X. and Goltz, G. (2000). Agent-based information flow for process industries' supply chain modelling, *Computers and Chemical Engineering* **24**: 1135–1141.

Geoffrion, A. (1970). Primal resource-directive approaches for optimizing nonlinear decomposable systems, *Operations Research* **18**: 375–403.

Geoffrion, A. (1972). Generalized benders decomposition, *Journal of Optimization Theory and Applications* **10**(4): 237–260.

Geoffrion, A. and Graves, G. (1974). Multicommodity distribution system design by benders decomposition, *Management Science* **20**(5): 822–844.

Giret, A. and Botti, V. (2004). Holons and agents, *Journal of Intelligent Manufacturing* **15**: 645–659.

Giulietti, F., Pollini, L. and Innocenti, M. (2000). Autonomous formation flight, *IEEE Control Systems Magazine* pp. 34–44.

Gjerdrum, J., Shah, N. and Papageorgiou, L. (2001). A combined optimization and agent-based approach to supply chain modelling and performance assessment, *Production Planning and Control* **12**(1): 81–88.

Glassey, C. (1973). Nested decomposition and multi-stage linear programs, *Management Science* **20**(3): 282–292.

Gou, L., Luh, P. and Kyoya, Y. (1998). Holonic manufactring scheduling: Architecture, cooperation, mechanism, and implementation, *Computers in Industry* **37**: 213–231.

Green, R. and Newbery, D. (1992). Competition in the british electric spot market, *Journal of Political Economy* **100**: 929–953.

Grossmann, I. and Daichendt, M. (1996). New trends in optimization-based approaches to process synthesis, *Computers and Chemical Engineering* **20**(6-7): 665–683.

Grothey, A. (2001). *Decomposition Methods for Nonlinear Nonconvex Optimization Problems*, Phd thesis, Department of Mathematics and Statistics, University of Edinburgh, Edinburgh, UK.

Grothey, A., Leyffer, S. and McKinnon, K. (1999). A note on feasibility in benders decomposition, *Technical Report NA-188*, University of Dundee, Dundee, UK.

Guo, Y., Hill, D. and Wang, Y. (2000). Nonlinear decentralized control of large-scale power systems, *Automatica* **36**: 1275–1289.

Heikkilä, T., Järviluoma, M. and Juntunen, T. (1997). Holonic control for manufacturing systems: Functional design of a manufacturing robot cell, *Integrated Computer-Aided Engineering* **4**: 202–218.

Ho, J. and Manne, A. (1974). Nested decomposition for dynamic models, *Mathematical Programming* **6**: 121–140.

Hobbs, B., Metzler, C. and Pang, J.-S. (2000). Strategic gaming analysis for electricity power systems: An mpec approach, IEEE *Transactions on Power Systems* **15**(2): 638–645.

Hou, Z.-G. (2001). A hierarchical optimization neural network for large-scale dynamic systems, *Automatica* **37**: 1931–1940.

Jackson, R. (1964a). A generalized variational treatment of optimization problems in complex chemical plants, *Chemical Engineering Science* **19**: 253–260.

Jackson, R. (1964b). Some algebraic properties of optimization problems in complex chemical plants, *Chemical Engineering Science* **19**: 19–31.

Jamshidi, M. (1983). *Large Scale Systems, Modelling and Control*, North Holland, New York, USA.

Jennings, N. and Bussmann, S. (2003). Agent-based control systems, *IEEE Control Systems Magazine* **23**(3): 61–74.

Jennings, N., Faratin, P., Norman, T., O'Brien, P., Odgers, B. and Alty, J. (2000). Implementing a business process management system using ADEPT: A real-world case study, *International Journal of Applied Artificial Intelligence* **14**(5): 421–465.

Johari, R. and Tan, D. (2001). End-to-end congestion control for the internet: Delays and stability, *IEEE/ACM Transactions on Networking* **9**(6): 818–832.

Jones, A. and McLean, C. (1986). A proposed hierarchical control model for automated manufacturing systems, *Journal of Manufacturing Systems* **5**(1): 15–25.

Jose, R. A. and Ungar, L. H. (2000). Pricing interprocess streams using slack auctions, *AIChE Journal* **46**(3): 575–587.

Julka, N., Karimi, I. and Srinivasan, R. (2002). Agent-based supply chain management—2: a refinery application, *Computers and Chemical Engineering* **26**: 1771–1781.

Julka, N., Srinivasan, R. and Karimi, I. (2002). Agent-based supply chain management—1: Framework, *Computers and Chemical Engineering* **26**: 1755–1769.

Katebi, M. and Johnson, M. (1997). Predictive control design for large-scale systems, *Automatica* **33**(3): 421–425.

Keller, G. E. and Bryan, P. F. (2000). Process engineering: Moving in new directions, *Chemical Engineering Progress* **96**(1): 41–50.

Kelly, F., Maulloo, A. and Tan, D. (1998). Rate control in communication networks: Shadow prices, proportional fairness, and stability, *Journal of the Operational Research Society* **49**: 237–252.

Keviczky, T., Borrelli, F. and Balas, G. (2006). Decentralized receding horizon control for large scale dynamically decoupled systems, *Automatica* **42**: 2105–2115.

Kisala, T., Trevino-Lozano, R., Boston, J., Britt, H. and Evans, L. (1987). Sequential modular and simultaneous modular strategies for process flowsheet optimization, *Computers and Chemical Engineering* **11**(6): 567–579.

Kleindorfer, P., Wu, D.-J. and Fernando, C. (2001). Strategic gaming in electric power markets, *European Journal of Operational Research* **130**: 156–168.

Koestler, A. (1967). *The Ghost in the Machine*, Hutchinson and Co., London, UK.

Koolen, J. (1998). Simple and robust design of chemical plants, *Computers and Chemical Engineering* **22**(Suppl.): S255–S262.

Korilis, Y. and Lazar, A. (1995). On the existence of equilibria in noncooperative optimal flow control, *Journal of the* ACM **42**: 584–613.

Korilis, Y., Lazar, A. and Orda, A. (1997). Capacity allocation under noncooperative routing, *IEEE Transactions on Automatic Control* **42**(3): 309–325.

Kornai, J. and Liptak, T. (1965). Two-level planning, *Econometrica* **33**(1): 141–169.

Landau, R. and Arora, A. (1999). The chemical industry: From the 1850s until today, *Business Economics* pp. 7–15.

Lasdon, L. S. (1970). *Optimization Theory for Large Systems*, The Macmillan Company, London.

Leitão, P. and Restivo, F. (2006). Adacor: A holonic architecture for agile and adaptive manufacturing control, *Computers in Industry* **57**: 121–130.

Lenhoff, A. and Morari, M. (1982). Design of resilient processing plants - i, *Chemical Engineering Science* **37**(2): 245–258.

Liu, J.-S. and Sycara, K. (1997). Coordination of multiple agents for production environment, *Annals of Operations Research* **75**: 235–289.

Lu, Z. (2003). Challenging control problems and emerging technologies in enterprise optimization, *Control Engineering Practice* **11**: 847–858.

Mah, R. (1990). *Chemical Process Structures and Information Flows*, Butterworth Series in Chemical Engineering, Butterworth-Heinemann, Boston, USA.

Mahey, P., Benchakroun, A. and Boyer, F. (2001). Capacity and flow assignment of data networks by generalized benders decomposition, *Journal of Global Optimization* **20**: 173–193.

Maloni, M. and Benton, W. (1997). Supply chain parternships: Opportunities for operations research, *European Journal of Operatioanl Research* **101**: 419–429.

Maturana, F. P. and Norrie, D. (1996). Multi-agent mediator architecture for distributed manufacturing, *Journal of Intelligent Manufacturing* **7**: 257–270.

Maturana, F. P., Tichý, P., Slechta, P., Staron, R., Discenzo, F. and Hall, K. (2003). *Multi-agent Systems III - Proceedings of CEEMAS 2003, Prague*, Vol. 2691 of *Lecture Notes in Artificial Intelligence*, Springer-Verlag, Berlin, chapter A Highly Distributed Intelligent Multi-agent Architecture for Industrial Automation, pp. 522–532.

Mařik, V., Fletcher, M. and Pěchouček (2002). *Multi-Agent-Systems and Applications II*, Vol. 2322 of *Lecture Notes in Computer Science*, Springer-Verlag, Berlin, chapter Holons and Agents: Recent Developments and Mutual Impacts, pp. 233–267.

Mařik, V. and McFarlane, D. (2005). Industrial adoption of agent-based technologies, *IEEE Intelligent Systems Magazine* **20**(1): 27–35.

McFarlane, D. C. (1995). Holonic manufacturing systems in continuous processing: Concepts and control requirements, *Proceedings of ASI'95* .

McFarlane, D. C. and Bussmann, S. (2000). Developments in holonic production planning and control, *Production Planning and Control* **11**(6): 522–536.

Mesarovic, M. D., Macko, D. and Takahara, Y. (1970). *Theory of Hierarchical Multilevel, Systems*, Academic Press, New York.

Molina, F. (1979). A survey of resource directive decomposition in mathematical programming, *Computing Surveys* **11**(2): 95–104.

Morari, M., Arkun, Y. and Stephanopoulos, G. (1980). Studies in the synthesis of control structures for chemical processes: Part i, *AIChE Journal* **26**(2).

Niemand, M. (2003). *Assessing the suitability of holonic control to the commodity petrochemical industry*, Master's, Faculty of Engineering, University of Pretoria, South Africa.

O'Brien, L. and Woll, D. (2005). DCS marks 30 year journey to operational excellence, *ARC Insights Report 2005-36MP*, ARC, Allied Drive, Dedham, MA 02026, USA, www.arcweb.com.

O'Neill, R. (1976). Nested decomposition of multistage convex programs, *SIAM Journal of Control and Optimization* **14**(3): 409–418.

Orda, A., Rom, R. and Shimkin, N. (1993). Competitive routing in multiuser communication networks, *IEEE/ACM Transactions on Networking* **1**(5): 510–521.

Papageorgiou, L. and Pantelides, C. (1996). Optimal campaign planning/scheduling of multipurpose batch/semicontinous plants. 1. mathematical formulation 2. mathematical decomposition, *Industrial and Engineering Chemistry Research* **35**: 488–529.

Perea-López, E., Grossmann, I., Ydstie, B. and Tahmassebi, T. (2000). Dynamic modeling and classical control theory for supply chain management, *Computers and Chemical Engineering* **24**: 1143–1149.

Perea-López, E., Grossmann, I., Ydstie, B. and Tahmassebi, T. (2001). Dynamic modelling and decentralized control of supply chains, *Industrial and Engineering Chemistry Research* **40**: 3369–3383.

Pěchouček, M. and Mařik, V. (2006). Review of industrial deployment of multi-agent systems, *Submitted to Journal of Autonomous Agents and Multi-Agent Systems*

Rannanjärvi, L. and Heikkilä, T. (1998). Software development for holonic manufacturing systems, *Computers in Industry* **37**: 233–253.

Rockafellar, R. (1970). *Convex Analysis*, Vol. 28 of *Princeton Mathematical Seires*, Princeton University Press, Princeton, New Jersey.

Rudd, D. and Watson, C. (1968). *Strategy of Process Engineering*, John Wiley, New York, USA.

Sahinidis, N. and Grossmann, I. (1991). Convergence properties of generalized benders decomposition, *Computers and Chemical Engineering* **15**(7): 481–491.

Samad, T., McLaughlin, P. and Lu, J. (2007). System architecture for process automation: Review and trends, *Journal of Process Control* **17**: 191–201.

Samyudia, Y., Lee, P. and Cameron, I. (1994). A methodology for multi-unit control design, *Chemical Engineering Science* **49**(23): 3871–3882.

Samyudia, Y., Lee, P., Cameron, I. and Green, M. (1995). A new approach to decentralised control design, *Chemical Engineering Science* **50**(11): 1695–1706.

Sandell, N. R., Varaiya, P., Athans, M. and Safonov, M. (1978). Survey of decentralized control methods for large scale systems, *IEEE Transactions on Automatic Control* **AC-23**(2): 108–128.

Schug, B. and Realff, M. (1996). Design of standardized, modular, chemical processes, *Computers and Chemical Engineering* **20**: S435–S441.

Seidel, D. (1994). HMS - Strategies, vol 0: Overview, *Technical report*, IMS/HMS TC5 Deliverables, HMS - Holonic Manufacturing Systems.

Seilonen, I., Appelqvist, P., Halme, A. and Koskinen, K. (2002). Agent-based approach to fault-tolerance in proess automation system, *Proc. of 3rd International Symposium on Robotics and Automation, ISRA'02*, Toluca, Mexico.

Seilonen, I., Appelqvist, P., Vainio, M., Halme, A. and Koskinen, K. (2002). A concept of an agent-augmented process automation system, *Proc. of 17th International Symposium on Intelligent Control, ISIC'02*, Vancouver, Canada.

Senehi, M., Wallace, S. and Luce, M. (1992). An architecture for manufacturing systems integration, *Proceedings of the Manufacturing International Conference*, Dallas, TX, USA.

Shah, N. (2004). Pharmaceutical supply chains: Key issues and strategies for optimisation, *Computers and Chemical Engineering* pp. 929–941.

Shah, N. (2005). Process industry supply chains: Advances and challenges, *Computers and Chemical Engineering* 29: 1225–1235.

Shen, W., Hao, Q., Yoon, H. and Norrie, D. (2006). Applications of agent-based systems in intelligent manufacturing: An updated review, *Advanced Engineering Informatics* 20: 415–431.

Shen, W., Maturanan, F. and Norrie, D. (2000). MetaMorph II: An agent-based architecture for distributed intelligent design and manufacturing, *Journal of Intelligent Manufacturing* 11: 237–251.

Shen, W., Wang, L. and Hao, Q. (2006). Agent-based distributed manufacturing process planning and scheduling: A state-of-the-art survey, *IEEE Transactions on Systems, Man and Cybernetics - Part C: Applications and Reviews* 36(4): 563–577.

Shobrys, D. and White, D. (2002). Planning, scheduling and control systems: Why they cannot work together, *Computers and Chemical Engineering* 26: 149–160.

Smith, C. (1970). Digital control of industrial processes, ACM *Computer Surveys* 2(3): 211–241.

Smith, R. G. (1980). The contract net protocol: High-level communicatoin and control in a distributed problem solver, *IEEE Transactions on Computers* C-29(12): 1104–1113.

Sousa, P. and Ramos, C. (1998). A dynamic scheduling holon for manufacturing orders, *Journal of Intelligent Manufacturing* 9: 107–112.

Stothert, A. and MacLeod, I. (2000). Competitive bidding as a control problem, *IEEE Transactions on Power Systems* 15(1): 88–94.

Strader, T., Lin, F.-R. and Shaw, M. (1998). Information infrastructure for electronic virtual organization management, *Decision Support Systems* 23: 75–94.

Suda, H. (1989). Future factory system formulated in japan, *Japanese Journal of Advanced Automation Technology* 1: 67–76.

Suda, H. (1990). Future factory system formulated in japan - part 2, *Japanese Journal of Advanced Automation Technology* 2(1): 58–66.

Takama, N. and Umeda, T. (1980). Multi-level, multi-objective optimization in process engineering, *Chemical Engineering Science* 36: 129–136.

Tayur, S., Ganeshan, R. and Magazine, M. (1999). *Quantitative Models for Supply Chain Management*, Kluwer's International Series in Operations Research and Management Science, Kluwer Academics, MA, USA.

Tenney, R. and Sandell, N. (1981). Strategies for distributed decisionmaking, *IEEE Transactions on Sytems, Man, and Cybernetics* SCM-11(8): 527–538.

ANSI/ISA (2003). Isa s95 - enterprise control system integration part 1: Models and terminology, *International standard*, Instrumentation, Systems and Automation Society, USA.

Tharumarajah, A. (2001). Survey of resource allocation methods for distributed manufacturing systems, *Production Planning and Control* **12**(1): 58–68.

Tharumarajah, A. and Wells, A. (1996). A behaviour-based approach to scheduling in distributed manufacturing systems, *Journal of Computer Aided Engineering, Special issue on Intelligent Manfacturing Systems* .

Tharumarajah, A., Wells, J. and Nemes, L. (1996). Comparison of bionic, fractal and holonic manufacturing system concepts, *Internationla Journal of Computer Integrated Manufacturing* **9**(3): 217–226.

Tousain, R. (2002). *Dynamic Optimization in Business-Wide Process Control*, IOS Press, Amsterdam, The Netherlands.

Tousain, R. and Bosgra, O. (2006). Market-oriented scheduling and economic optimization of continuous multi-grade chemical processes, *Journal of Process Control* **16**(3): 291–302.

Ueda, K. (1992). An approach to bionic manufacturiing systems based on dna-type information, *Proceedings of ICOOMS'92*.

Valckenaers, P. and van Brussel, H. (2005). *Proc. of HoloMAS 2005, V. Mark, R.W. Brennan, M. Pechoucek (Eds.)*, Lecture Notes in Artificial Intelligence 3593, Springer-Verlag, Berlin, chapter Fundamental Insights into Holonic Systems Design, pp. 11–22.

Valckenaers, P., van Brussel, H., Kollingbaum, M. and Bochmann, O. (2001). Multi-agent coordination and control using stigmergy applied to manufacturing control, *Proceedings of ACAI 2001, Also as Lecture Notes in Artificial Intelligence, 2086, Springer-Verlag, Berlin, Germany*.

van Brussel, H. V., Bongaerts, L., Wyns, J., Valckenaers, P. and Ginderachter, T. (1999). A conceptual framework for holonic manufacturing: Identification of manufacturing holons, *Journal of Manufacturing Systems* **18**(1): 35–52.

van Brussel, H., Wyns, J., Valckenaers, P., Bongaerts, L. and Peeters, P. (1998). Reference architecture for holonic manufacturing systems: PROSA, *Computers in Industry* **37**: 255–274.

Váncza, J. and Márkus (2000). An agent model for incentive-based production scheduling, *Computers in Industry* **43**: 173–187.

Váncza, J. and Márkus, A. (1998). Holonic manufacturing with economic rationality, *Proc. of IMS-EUROPE Workshop, Lausanne, Switzerland*, pp. 383–394.

Venkat, A., Hiskens, I., Rawlings, J. and Wright, S. (2006). Distributed mpc strategies with application to power system automatic generation control, *Technical Report 2006-05*, Department of Chemical and Biological Engineering , University of Wisconsin, Madison-53706, USA.

Vinnicombe, G. (2000). On the stability of end-to-end congestion control for the internet, *Technical Report CUED/F-INFENG/TR.398*, Department of Engineering, University of Cambridge, Cambridge, UK.

Šiljak, D. (1991). *Decentralized Control of Complex Systems*, Vol. 184 of *Mathematics in Science and Engineering*, Academic Press, Inc., Boston, USA.

Westerberg, A., Hutchison, H., Motard, R. and Winter, P. (1979). *Process Flowsheeting*, Cambridge University Press, Cambridge, UK.

Williams, T. (1989). *A Reference Model for Computer Integrated Manufacturing(CIM): A Description from the Viewpoint of Industrial Automation*, Instrumentation, Systems and Automation Society, North Carolina, USA.

Wittrock, R. (1985). Dual nested decomposition of staircase linear programs, *Mathematical Programming Study* **24**: 65–86.

Wooldridge, M. (2002). *An Introduction to Multiagent Systems*, John Wiley & Sons., Chichester, UK.

Wu, P., Hartman, J. and Wilson, G. (2003). A demand-shifting feasibility algorithm for benders decomposition, *European Journal of Operational Research* **148**: 570–583.

Wyns, J. (1999). *Reference Architecture for Holonic Manuafacturing Systems - the key to Support to Evolution and Reconfiguration*, Phd thesis, Department Werktuigkunde Afdeling Productietechnieken, Katholieke Universiteit, Leuven, Belgium.

Zwegers, A. (1998). *On systems architecting: A Study in Shop-floor Control to Determine Architecting Concepts and Principles*, Phd thesis, Technische Universiteit Eindhoven, Eindhoven, Belgium.

Index